Lecture Notes in Mathematics

A collection of informal reports and seminars

Edited by A. Dold, Heidelberg and B. Eckmann, Zürich

T0240602

251

The Theory of Arithmetic Functions

Proceedings of the Conference at Western Michigan University, April 29 – May 1, 1971

Edited by Anthony A. Gioia and Donald L. Goldsmith
Western Michigan University, Kalamazoo, MI/USA

Springer-Verlag
Berlin · Heidelberg · New York 1972

AMS Subject Classifications (1970): 10-xx, 10-02, 10A xx, 10H xx, 10K xx

ISBN 3-540-05723-4 Springer-Verlag Berlin · Heidelberg · New York
ISBN 0-387-05723-4 Springer-Verlag New York · Heidelberg · Berlin

Offsetdruck: Julius Beltz, Hemsbach/Bergstr.

FOREWORD

This book contains the proceedings of the Conference on the Theory of Arithmetic Functions, held at Western Michigan University, Kalamazoo, Michigan from April 29 to May 1, 1971. Except for slight changes in format, the manuscripts of the lectures are published here as supplied to us by the speakers.

Our objective in arranging the Conference was to bring together as many workers in the field of Arithmetic Functions as possible. Because of limitations on the time available for the Conference, it was not possible to include all of the material which was submitted. We regret the omission from the program of several excellent papers.

We wish to thank the following for contributing to the success of the Conference:

The speakers and other participants;

The National Science Foundation, for financial support;

Western Michigan University, for financial support and use of facilities;

The Visiting Scholars Program, Western Michigan University, for financial support;

The Honors College, Western Michigan University, for financial support;

Professor Leonard Carlitz, for the Visiting Scholar's Lecture;

Professor A. Bruce Clarke, Chairman of the Department of Mathematics, for advice and support;

Professor Yousef Alavi, for assistance in organization;

Mrs. Patricia Williams, for administrative assistance;

Graduate assistants Mr. Timothy Carroll and Mr. Arnold Veldkamp;

Undergraduate assistants Miss Alexis Kaczynski and Miss Susan O'Brien;

In addition, we thank the following for their part in the publication of this book:

Springer-Verlag, the publishers;

The authors, for their cooperation and promptness;

Mrs. Darlene Lard, who typed the manuscript with exceptional patience and skill;

Miss Kaczynski and Miss O'Brien, who helped with the proofreading.

Anthony A. Gioia

Donald L. Goldsmith

TABLE OF CONTENTS

GEORGE E. ANDREWS

Sieves for theorems of Euler, Rogers, and Ramanujan 1

BRUCE C. BERNDT

The Voronoï summation formula 21

LEON BERNSTEIN

The arithmetic functions of the Jacobi-Perron algorithm 37

L. CARLITZ

Eulerian numbers and operators 65

L. CARLITZ, V. E. HOGGATT, and RICHARD SCOVILLE

Some functions related to Fibonacci and Lucas representations 71

L. M. CHAWLA and JOHN E. MAXFIELD

On the classification and evaluation of some partition functions and their graphs 103

HAROLD G. DIAMOND

Two oscillation theorems 113

P. ERDÖS and M. V. SUBBARAO

On the iterates of some arithmetic functions 119

JANOS GALAMBOS

Distribution of additive and multiplicative functions 127

EMIL GROSSWALD

Oscillation theorems 141

D. J. LEWIS

Curves with abnormally many integral points 169

K. NAGESWARA RAO

On certain arithmetical sums 181

C. RYAVEC

The use of measure-theoretic methods
in the study of additive arithmetic functions 193

DAVID A. SMITH

Generalized arithmetic function algebras 205

M. V. SUBBARAO

On some arithmetic convolutions 247

D. SURYANARAYANA

The number of bi-unitary divisors
of an integer 273

CHARLES R. WALL

Density bounds for the sum of
divisors function 283

SIEVES FOR THEOREMS OF EULER, ROGERS, AND RAMANUJAN*

George E. Andrews, Massachusetts Institute of Technology
and Pennsylvania State University

1. Introduction. The theory of sieves has become an important and sophisticated tool in the theory of numbers during the past 52 years. Many important and deep results have been proved related to such unanswered questions as the Twin Primes Problem and Goldbach's Conjecture. For the most part sieves have been employed in problems concerning primes. The object of this paper is to introduce a sieve related to the theory of partitions.

The sieve method may be thought of as breaking into two parts. First the principle of inclusion-exclusion (or some similar combinatorial counting process) is applied to obtain an expression for a particular arithmetic function. Then the new expression is treated number-theoretically to yield information about the original function. The most elementary example of this is the simple sieve estimate for $\pi(x)$, the number of primes not exceeding x. If $p_1, p_2, \ldots, p_r$ denote the first r primes, $A(x,r)$ denotes the number of integers not exceeding x that are not divisible by any of $p_1, p_2, \ldots, p_r$, and $[x]$ denotes the largest integer not exceeding x, then by the principle of inclusion-exclusion [LeVeque; 8; p. 98]

$$A(x,r) = [x] - \sum_{i=1}^{r} \left[\frac{x}{p_i}\right] + \sum_{1 \le i < j \le r} \left[\frac{x}{p_i p_j}\right] - \cdots + (-1)^r \left[\frac{x}{p_1 \cdots p_r}\right].$$

The estimate $\pi(x) = O(x/\log\log x)$ is then obtained by observing that $\pi(x) \le A(x,r) + r$ and estimating $A(x,r)$ by carefully comparing it with

*This research was partially supported by National Science Foundation Grant GP-9660.

$$x - \sum_{i=1}^{r} \frac{x}{p_i} + \sum_{1 \le i < j \le r} \frac{x}{p_i p_j} - \cdots + (-1)^r \frac{x}{p_1 \cdots p_r}$$

$$= x \prod_{i=1}^{r} \left(1 - \frac{1}{p_i}\right).$$

We propose to examine a new technique in the theory of partitions that is at least a cousin of the sieves of prime number theory. This technique involves an "inclusion-exclusion" type enumeration of certain partitions; then one makes a careful analysis of the expressions obtained.

We propose to give sieve-theoretic proofs of the following three theorems from additive number theory

(1.1) $$1 = \frac{1}{(q)_\infty} - \frac{q}{(q)_\infty} - \frac{q^2}{(q)_\infty} + \frac{q^5}{(q)_\infty} + \frac{q^7}{(q)_\infty} - \cdots,$$

(1.2) $$\sum_{n=0}^{\infty} \frac{q^{n^2}}{(q)_n} = \frac{1}{(q)_\infty} - \frac{q^2}{(q)_\infty} - \frac{q^3}{(q)_\infty} + \frac{q^9}{(q)_\infty} + \frac{q^{11}}{(q)_\infty} - \cdots$$

$$= \frac{1}{(q)_\infty} + \sum_{n=1}^{\infty} (-1)^n \left\{\frac{q^{\frac{1}{2}n(5n-1)}}{(q)_\infty} + \frac{q^{\frac{1}{2}n(5n+1)}}{(q)_\infty}\right\},$$

(1.3) $$\sum_{n=0}^{\infty} \frac{q^{n^2+n}}{(q)_n} = \frac{1}{(q)_\infty} - \frac{q}{(q)_\infty} - \frac{q^4}{(q)_\infty} + \frac{q^7}{(q)_\infty} + \frac{q^{13}}{(q)_\infty} - \cdots$$

$$= \frac{1}{(q)_\infty} + \sum_{n=1}^{\infty} (-1)^n \left\{\frac{q^{\frac{1}{2}n(5n-3)}}{(q)_\infty} + \frac{q^{\frac{1}{2}n(5n+3)}}{(q)_\infty}\right\},$$

where

$$(a;q)_n = (a)_n = (1-a)(1-aq)\cdots(1-aq^{n-1})$$

and

$$(a;q)_\infty = (a)_\infty = \lim_{n\to\infty} (a;q)_n.$$

The identity (1.1) is essentially Euler's Pentagonal Number Theorem [Hardy and Wright; 7; p. 284] while (1.2) and (1.3) are known as the Rogers-Ramanujan identities [Hardy and Wright; 7; p. 290].

The sieves that we shall apply act on the set of all partitions of an integer, each partition being classified according to its "successive ranks" (see Section 2). The sieve related to (1.1) sieves out all partitions except the empty partition, whence the 1 on the left-hand side. The sieves for (1.2) and (1.3) remove all partitions except those related in an obvious way to the left-hand series.

In Sections 3 and 4, we shall treat (1.1) in detail. Section 5 will be devoted to an outline of the proofs of (1.2), (1.3), and a description of other results.

2. <u>Successive ranks</u>. We shall consider partitions of an integer n into positive integral parts, where the order of the parts is irrelevant. To facilitate matters we shall represent each partition with the parts in decreasing order of magnitude. We shall also be interested in the graphical representation of partitions. For example, the partition π of 15 given by $5 + 4 + 4 + 2$ is represented graphically as follows.

```
. . . . .
. . . .
. . . .
. .
```

The number of nodes in the j-th row of the graph equals the j-th part of π. To obtain what is called the conjugate partition $C\pi$, we read the nodes vertically; in this instance $C\pi$ is $4 + 4 + 3 + 3 + 1$.

In 1944, F. J. Dyson [5] introduced an important parameter of a partition called its <u>rank</u>, the largest part minus the number of parts. Thus the rank of the partition $5 + 4 + 4 + 2$ is $5 - 4 = 1$. Dyson conjectured and Atkin and Swinnerton-Dyer [3] proved the following result connecting the rank with the Ramanujan congruences:

Let $K(r,a;n)$ denote the number of partitions of n in which the rank is congruent to a modulo r; then for

$0 \le a \le 4$, $K(5,a;5n + 4) = \frac{1}{5}\, p(5m + 4)$ and for $0 \le a \le 6$, $K(7,a;7n + 5) = \frac{1}{7}\, p(7n + 5)$.

From these results, we see directly that

$$p(5n + 4) \equiv 0 \pmod{5},$$

and

$$p(7n + 5) \equiv 0 \pmod{7}.$$

In a later paper, Atkin [4] introduced what he termed the successive ranks. He was motivated to study these further ranks in an unsuccessful search for a "rank-theoretic" interpretation of the further Ramanujan congruence:

$$p(11n + 6) \equiv 0 \pmod{11}.$$

Definition 1. If π denotes a partition, then $d_r(\pi)$, the r-rank of π, is defined as the difference between the r-th part of π and the r-th part of the conjugate of π; furthermore, the r-rank is defined only if both the r-th row and r-th column of π have a node in common.

Computing the successive ranks of a partition is done most easily by means of the graphical representation. Thus for the partition π: $15 = 5 + 4 + 4 + 2$,

the ranks are $d_1(\pi) = 1$, $d_2(\pi) = 0$, $d_3(\pi) = 1$. The computation is done by subtracting the number of vertical nodes in the r-th indicated right angle from the number of horizontal nodes. We see from the graphical representation that if the Durfee square of π (see [Hardy and Wright; 7; p. 281]) is of side s, then there are exactly s successive ranks for π.

Our interest lies in the "oscillation" of the successive ranks between positive and nonpositive values.

<u>Definition</u> 2. Let h be the largest integer for which there exists a sequence $j_1 < j_2 < \cdots < j_h$ such that

$$d_{j_1}(\pi) > 0, \quad d_{j_2}(\pi) \le 0, \quad d_{j_3}(\pi) > 0, \quad d_{j_4}(\pi) \le 0,$$

and so on (more precisely $(-1)^i(\frac{1}{2} - d_{j_i}(\pi)) > 0$ for $1 \le i \le h$). We define h to be the <u>positive oscillation</u> of π.

<u>Definition</u> 3. Let k be the largest integer for which there exists a sequence $j_1 < j_2 < \cdots < j_k$ such that

$$d_{j_1}(\pi) \le 0, \quad d_{j_2}(\pi) > 0, \quad d_{j_3}(\pi) \le 0, \quad d_{j_4}(\pi) > 0,$$

and so on (more precisely $(-1)^i(d_{j_i}(\pi) - \frac{1}{2}) > 0$ for $1 \le i \le k$). We define k to be the <u>negative oscillation</u> of π.

For example, let us consider the partition π given by

$$112 = 17 + 15 + 10 + 10 + 10 + 10 + 8 + 8 + 7 + 5 + 5 + 4 + 3.$$

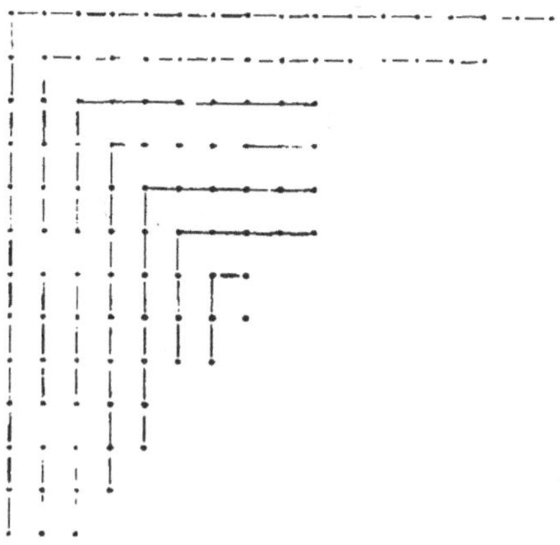

$d_1(\pi) = 4$, $d_2(\pi) = 2$, $d_3(\pi) = -3$, $d_4(\pi) = -2$, $d_5(\pi) = -1$, $d_6(\pi) = 1$, $d_7(\pi) = -1$, $d_8(\pi) = 0$. Consequently the positive oscillation of π is 4 (e.g., choose $j_1 = 1$, $j_2 = 3$, $j_3 = 6$, $j_4 = 7$), and the negative oscillation of π is 3 (e.g., choose $j_1 = 3$, $j_2 = 6$, $j_3 = 7$).

Proposition 1. The positive oscillation of π is 1 larger than the negative oscillation if $d_1(\pi) > 0$ and is 1 less if $d_1(\pi) \le 0$.

Proof. Suppose that h denotes the positive oscillation of π and k the negative oscillation. If $d_1(\pi) > 0$, let $j_1 < j_2 < \cdots < j_k$ be the sequence described in Definition 3. Then the sequence $j_1' < j_2' < \cdots < j_k' < j_{k+1}'$ given by $j_1' = 1$, $j_i' = j_{i-1}$ for $2 \le i \le k+1$ is of the type described in Definition 2; consequently $h \ge k+1$. However it is obvious that $k \ge h-1$ because in the maximal sequence $j_1^* < j_2^* < \cdots < j_h^*$ defining h (see Definition 2) we see that $j_1'' < j_2'' < \cdots < j_{h-1}''$ (where $j_i'' = j_{i+1}^*$) is of the type described in Definition 3. Therefore if $d_1(\pi) > 0$, then $h = k+1$.

If $d_1(\pi) \le 0$, a mirror image of the above argument shows that $k = h+1$. □

In the next section we shall develop a sieve related to successive ranks.

3. Generating functions for partitions with prescribed oscillation. We shall study certain functions related to the positive and negative oscillation of a partition.

Definition 4. Let $p(a,b;\mu;N)$ (resp., $m(a,b;\mu;N)$) denote the number of partitions of N with at most b parts, with largest part at most a and with positive (resp., negative) oscillation at least μ.

To these partition functions we associate the related generating functions:

$$p(a,b;\mu;q) = \sum_{N \ge 0} p(a,b;\mu;N)q^N, \tag{3.1}$$

$$M(a,b;\mu;q) = \sum_{N \ge 0} m(a,b;\mu;N)q^N. \tag{3.2}$$

Lemma 1. <u>The following recurrence relations hold for</u> $\mu \geq 1$:

$$m(a,b;\mu;N) - m(a-1,b;\mu;N) - m(a,b-1;\mu;N) + m(a-1,b-1;\mu;N)$$

$$(3.3)\qquad = \begin{cases} m(a-1,b-1;\mu;N-a-b+1), & \text{if } a > b, \\ p(a-1,b-1;\mu-1;N-a-b+1), & \text{if } a \leq b; \end{cases}$$

$$p(a,b;\mu;N) - p(a-1,b;\mu;N) - p(a,b-1;\mu;N) + p(a-1,b-1;\mu;N)$$

$$(3.4)\qquad = \begin{cases} m(a-1,b-1;\mu-1;N-a-b+1), & \text{if } a > b, \\ p(a-1,b-1,\mu;N-a-b+1), & \text{if } a \leq b. \end{cases}$$

<u>Proof</u>. Let us start by examining the left side of (3.3). The expression $m(a,b;\mu;N) - m(a-1,b;\mu;N)$ denotes the number of partitions of N with at most b parts, with negative oscillation at least μ, and with largest part <u>equal</u> to a. Consequently, the expression $\{m(a,b;\mu;N) - m(a-1,b;\mu;N)\} - \{m(a,b-1;\mu;N) - m(a-1,b-1;\mu;N)\}$ (which is the left-hand side of (3.3)) denotes the number of partitions of N with exactly b parts, with negative osicllation at least μ, and with largest part <u>equal</u> to a. We now transform such partitions by deleting the largest part and subtracting 1 from each of the remaining parts; in this way, we are left with a partition of $N-a-b+1$ into at most $b-1$ parts with largest part at most $a-1$. Furthermore if $a > b$, then $d_1(\pi) > 0$; hence the removal of this outer angle of nodes from the Ferrars graph has no effect on the negative oscillation which is therefore still equal to μ. Thus our transformed partition is of the type enumerated by $m(a-1,b-1;\mu;N-a-b+1)$. Since the above procedure is clearly reversible, we see that

$$m(a,b;\mu;N) - m(a-1,b;\mu;N) - m(a,b-1;\mu;N) + m(a-1,b-1;\mu;N)$$
$$= m(a-1,b-1;\mu;N-a-b+1),$$

and so the top half of (3.3) is verified.

If $a \leq b$, then $d_1(\pi) \leq 0$, and removal of the outer angle of nodes from the Ferrars graph may indeed affect the negative oscillation. In fact, the resulting partition is now characterized by the fact that it has positive oscillation of at least $\mu-1$. The transformed partitions in this instance are of the type enumerated by

$p(a-1,b-1;\mu-1;N-a-b+1)$. Hence the lower half of (3.3) follows.

The proof of (3.4) is a mirror image of the proof for (3.3). $\square$

The results in Lemma 1 may now be translated directly into functional equations for the related generating functions. Hence, for $\mu \geq 1$,

$$M(a,b;\mu;q) - M(a-1,b;\mu;q) - M(a,b-1;\mu;q) + M(a-1,b-1;\mu;q)$$

$$(3.5) \qquad = q^{a+b-1} \begin{cases} M(a-1,b-1;\mu;q), & \text{if } a > b, \\ P(a-1,b-1;\mu-1;q), & \text{if } a \leq b; \end{cases}$$

$$P(a,b;\mu;q) - P(a-1,b;\mu;q) - P(a,b-1;\mu;q) + P(a-1,b-1;\mu;q)$$

$$(3.6) \qquad = q^{a+b-1} \begin{cases} M(a-1,b-1;\mu-1;q), & \text{if } a > b, \\ P(a-1,b-1;\mu;q), & \text{if } a \leq b. \end{cases}$$

Lemma 2. *The following relations hold:*

$$(3.7) \qquad P(a,b;0;q) = M(a,b;0;q) = \binom{a+b}{a}_q ;$$

for $\mu \geq 1$.

$$(3.8) \quad P(0,b;\mu;q) = P(a,0;\mu;q) = M(0,b;\mu;q) = M(a,0;\mu;q) = 0,$$

where

$$\binom{N}{M}_q = \begin{cases} \dfrac{(q)_N}{(q)_M \, (q)_{N-M}}, & \text{for } 0 \leq M \leq N, \\ 0, & \text{otherwise.} \end{cases}$$

Proof. The requirement that the negative (or positive) oscillation be at least 0 is clearly met by any partition. Therefore $P(a,b;0;q)$ and $M(a,b;0;q)$ are both the generating function for partitions with largest part at most a and at most b parts. It is well-known [Sylvester; 10; p. 269] that this generating function is

$$\binom{a+b}{a}_q .$$

Thus we have (3.7).

The requirement that either the largest part a (or the number of parts b) be at most 0 is not met by any partition save the empty partition, and the empty partition cannot have negative (or positive) oscillation $\mu \geq 1$. Hence (3.8) is established. □

Our next result might be described as the second part of the sieve. It will be the essential result after the "inclusion-exclusion" type procedure has been used.

Theorem 1. *If* $b = a$ *or* $a-1$, *then*

(3.9) $$M(a,b;2\mu;q) = q^{\mu(6\mu+1)}\binom{a+b}{b-3\mu}_q ,$$

(3.10) $$M(a,b;2\mu-1;q) = q^{(2\mu-1)(3\mu-2)}\binom{a+b}{b-3\mu+2}_q ,$$

(3.11) $$P(a,b;2\mu;q) = q^{\mu(6\mu-1)}\binom{a+b}{b+3\mu}_q ,$$

(3.12) $$P(a,b;2\mu-1;q) = q^{(2\mu-1)(3\mu-1)}\binom{a+b}{a+1-3\mu}_q .$$

Proof. We first note that equations (3.5), (3.6), (3.7), and (3.8) uniquely define $M(a,b;\mu;q)$ and $P(a,b;\mu;q)$ for each nonnegative integral a, b, and μ. We now consider the following functions:

(3.13) $$M^*(a,b;2\mu;q) = q^{\mu(6\mu+1)}\binom{a+b}{b-3\mu}_q , \quad \text{for } a \geq b, \ \mu \geq 1;$$

(3.14) $$M^*(a,b;2\mu-1;q) = q^{(2\mu-1)(3\mu-2)}\binom{a+b}{b-3\mu+2}_q , \quad \text{for } a \geq b, \ \mu \geq 1;$$

(3.15) $$P^*(a,b;2\mu;q) = q^{\mu(6\mu-1)}\binom{a+b}{b+3\mu}_q, \quad \text{for } b \ge a-1, \quad \mu \ge 1;$$

(3.16) $$P^*(a,b;2\mu-1;q) = q^{(2\mu-1)(3\mu-1)}\binom{a+b}{a+1-3\mu}_q, \quad \text{for } b \ge a-1, \quad \mu \ge 1;$$

(3.17) $$M^*(a-1,a+r;2\mu;q) = q^{6\mu^2+\mu}\frac{1-q^{(r+2)(a-3\mu)}}{1-q^{a-3\mu}}\binom{2a-2}{a-1-3\mu}_q + \sum_{j=0}^{r-1} q^{6\mu^2-2\mu+a+1+j}\frac{1-q^{(r-j)(a-3\mu)}}{1-q^{a-3\mu}}\binom{2a-2+j}{a-2-3\mu}_q,$$

for $a \ge 1$, $r \ge -1$, $\mu \ge 1$;

(3.18) $$M^*(a-1,a+r;2\mu-1;q) = q^{(2\mu-1)(3\mu-2)}\frac{1-q^{(r+2)(a-3\mu+2)}}{1-q^{a-3\mu+2}}\binom{2a-2}{a+3\mu-3}_q + \sum_{j=0}^{r-1} q^{6\mu^2-10\mu+a+5+j}\frac{1-q^{(r-j)(a-3\mu+2)}}{1-q^{a-3\mu+2}}\binom{2a-2+j}{a+3\mu-2+j}_q,$$

for $a \ge 1$, $r \ge -1$, $\mu \ge 1$;

(3.19) $$P^*(b+r,b-1;2\mu;q) = q^{6\mu^2-\mu}\frac{1-q^{(r+1)(b-3\mu+1)}}{1-q^{b-3\mu+1}}\binom{2b-1}{b+3\mu-1}_q + \sum_{j=0}^{r-2} q^{6\mu^2-4\mu+2+b+j}\frac{1-q^{(r-j-1)(b-3\mu+1)}}{1-q^{b-3\mu+1}}\binom{2b-1+j}{b+3\mu+j}_q,$$

for $b \ge 1$, $r \ge 0$, $\mu \ge 1$;

(3.20) $$P^*(b+r,b-1;2\mu-1;q) = q^{(2\mu-1)(3\mu-1)}\frac{1-q^{(r+1)(b-3\mu+2)}}{1-q^{b-3\mu+2}}\binom{2b-1}{b+1-3\mu}_q + \sum_{j=0}^{r-2} q^{(2\mu-1)(3\mu-1)+b-3\mu+3+j}\frac{1-q^{(r-1-j)(b-3\mu+2)}}{1-q^{b-3\mu+2}}\binom{2b-1+j}{b-3\mu}_q,$$

for $b \ge 1$, $r \ge 0$, $\mu \ge 1$;

(3.21) $$P^*(a,b;0;q) = M^*(a,b;0;q) = \binom{a+b}{a}_q.$$

We observe immediately that $P^*(a,b;\mu;q)$ and $M^*(a,b;\mu;q)$ satisfy (3.7) and (3.8). It is a matter of elementary (but lengthy) algebraic manipulation using the identities

$$(3.22)\qquad \binom{N}{M}_q = \binom{N-1}{M-1}_q + q^M \binom{N-1}{M}_q ,$$

$$(3.23)\qquad \binom{N}{M}_q = \binom{N-1}{M}_q + q^{N-M} \binom{N-1}{M-1}_q ,$$

and

$$(3.24)\qquad \binom{N}{M}_q = \binom{N}{N-M}_q ,$$

to prove that $P^*(a,b;\mu;q)$ and $M^*(a,b;\mu;q)$ also satisfy (3.5) and (3.6).

Recalling the first sentence of this proof, we thus see that for each nonnegative integral a, b, and μ,

$$(3.25)\qquad P^*(a,b;\mu;q) = P(a,b;\mu;q),$$

and

$$(3.26)\qquad M^*(a,b;\mu;q) = M(a,b;\mu;q).$$

The equations (3.9)-(3.12) now follow from (3.13) - (3.16), (3.25), and (3.26).

<u>Definition</u> 5. Let $p(\mu;N)$ (resp., $m(\mu;N)$) denote the number of partitions of N with positive (resp., negative) oscillation at least μ. We also let $P(\mu;q)$ and $M(\mu;q)$ denote the respective generating functions.

<u>Corollary</u>. For each $\mu \geq 0$,

$$(3.27)\qquad M(2\mu;q) = \frac{q^{\mu(3\mu+1)}}{(q)_\infty},$$

$$(3.28)\qquad M(2\mu-1;q) = \frac{q^{(2\mu-1)(3\mu-2)}}{(q)_\infty},$$

(3.29) $$P(2\mu;q) = \frac{q^{\mu(6\mu-1)}}{(q)_\infty},$$

(3.30) $$P(2\mu-1;q) = \frac{q^{(2\mu-1)(3\mu-1)}}{(q)_\infty}.$$

Proof. These results follow directly from the fact that for $|q| < 1$

$$|P(\mu;q) - P(a,a;\mu;q)| \le \sum_{n=a+1}^{\infty} p(n)|q|^n \to 0 \quad \text{as} \quad a \to \infty,$$

$$|M(\mu;q) - M(a,a;\mu;q)| \le \sum_{n=a+1}^{\infty} p(n)|q|^n \to 0 \quad \text{as} \quad a \to \infty,$$

and

$$\lim_{a\to\infty} q^X \begin{pmatrix} 2a+Y \\ a-Z \end{pmatrix}_q = \frac{q^X}{(q)_\infty}.$$

4. A sieve for Euler's Pentagonal Number Theorem (1.1). The following theorem may be thought of as the "inclusion-exclusion" aspect of the sieve we are constructing.

Theorem 2. For $n > 0$,

(4.1) $$p(0;n) + \sum_{\mu=1}^{\infty} (-1)^{\mu} m(\mu;n) + \sum_{\mu=1}^{\infty} (-1)^{\mu} p(\mu;n) = 0.$$

Proof. The above sum is a weighted count of partitions of n. First if π is a partition of n, $d_1(\pi) > 0$ and the positive oscillation of π is r, then by Proposition 1 the negative oscillation of π is $r-1$. Hence π is counted once by each of $p(0,n)$, $m(1;n)$, ..., $m(r-1,n)$, $p(1,n)$, ..., $p(r,n)$ and in the above sum the weight of the count is

$$1 + \sum_{\mu=1}^{r-1} (-1)^{\mu} + \sum_{\mu=1}^{r} (-1)^{\mu} = 1 + \sum_{\mu=1}^{r-1} (-1)^{\mu} - \sum_{\mu=0}^{r-1} (-1)^{\mu} = 1 - 1 = 0.$$

If $d_1(\pi) \le 0$, and the positive oscillation of π is r, then by Proposition 1 the negative oscillation of π is $r+1$. Hence the weighted count in this case is

$$1 + \sum_{\mu=1}^{r+1} (-1)^{\mu} + \sum_{\mu=1}^{r} (-1)^{\mu} = 1 - \sum_{\mu=0}^{r} (-1)^{\mu} + \sum_{\mu=1}^{r} (-1)^{\mu} = 1 - 1 = 0.$$

Thus each partition of n is counted with weight 0; that is,

$$p(0,n) + \sum_{\mu=1}^{\infty} (-1)^{\mu} m(\mu,n) + \sum_{\mu=1}^{\infty} (-1)^{\mu} p(\mu,n) = 0$$

for $n > 0$.

<u>Theorem</u> 3. (<u>Euler's Pentagonal Number Theorem (1.1)</u>).

$$1 = \frac{1}{(q)_\infty} + \sum_{n=1}^{\infty} (-1)^n q^{\frac{1}{2}n(3n-1)} \left\{\frac{1+q^n}{(q)_\infty}\right\}.$$

<u>Proof</u>. Multiply equation (4.1) by q^n, sum for $1 \le n < \infty$ and then add 1 to both sides of the equation.

$$1 = P(0;q) + \sum_{\mu=1}^{\infty} (-1)^{\mu} M(\mu;q) + \sum_{\mu=1}^{\infty} (-1)^{\mu} P(\mu;q)$$

$$= \frac{1}{(q)_\infty} - \sum_{\mu=1}^{\infty} M(2\mu-1;q) + \sum_{\mu=1}^{\infty} M(2\mu;q)$$

$$- \sum_{\mu=1}^{\infty} P(2\mu-1;q) + \sum_{\mu=1}^{\infty} P(2\mu;q)$$

$$= \frac{1}{(q)_\infty} - \sum_{\mu=1}^{\infty} \frac{q^{(2\mu-1)(3\mu-2)}}{(q)_\infty} + \sum_{\mu=1}^{\infty} \frac{q^{\mu(6\mu+1)}}{(q)_\infty}$$

$$= \frac{1}{(q)_\infty} - \sum_{\mu=1}^{\infty} \frac{q^{\frac{1}{2}(2\mu-1)(3(2\mu-1)-1)}(1+q^{2\mu-1})}{(q)_\infty}$$

$$+ \sum_{\mu=1}^{\infty} \frac{q^{\frac{1}{2}2\mu(3(2\mu)-1)}(1+q^{2\mu})}{(q)_\infty}$$

$$= \frac{1}{(q)_\infty} + \sum_{n=1}^{\infty} (-1)^n q^{\frac{1}{2}n(3n-1)} \left\{\frac{1+q^n}{(q)_\infty}\right\}. \quad \square$$

If we had used Theorem 1 directly for our sieve instead of the Corollary, we would have been able to obtain the following identity due to Schur [9].

$$\sum_{\lambda=-\infty}^{\infty} (-1)^\lambda q^{\frac{1}{2}\lambda(3\lambda+1)} \binom{n}{[\frac{1}{2}(n-3\lambda)]}_q = 1, \tag{4.2}$$

where $[x]$ denotes the largest integer not exceeding x. The case $a = b$ in Theorem 1 yields (4.2) with $n = 2a$, and $a = b-1$ in Theorem 1 yields (4.2) with $n = 2b-1$.

5. Sieve for the Rogers-Ramanujan identity (1.2) and (1.3). The technique developed in Sections 4 and 5 can be extended to provide an infinite family of sieves. We shall consider the extension that yields a proof of (1.2) and (1.3).

Definition 6. Let h be the largest integer for which there exists a sequence $j_1 < j_2 < \cdots < j_h$ such that

$$d_{j_1}(\pi) > 2k-a-1, \quad d_{j_2}(\pi) \le -(a-1), \quad d_{j_3}(\pi) > 2k-a-1, \quad d_{j_4}(\pi) \le -(a-1),$$

and so on. We define h to be the (k,a)-positive oscillation of π.

Definition 7. Let g be the largest integer for which there exists a sequence $j_1 < j_2 < \cdots < j_g$ such that

$$d_{j_1}(\pi) \le -(a-1), \quad d_{j_2}(\pi) > 2k-a-1, \quad d_{j_3}(\pi) \le -(a-1), \quad d_{j_4}(\pi) > 2k-a-1,$$

and so on. We define g to be the (k,a)-negative oscillation of π.

We remark that Definitions 2 and 3 correspond to the case $k = a = 1$.

Definition 8. Let $P_{k,a}(\mu;N)$ (resp., $M_{k,a}(\mu;N)$) denote the number of partitions of N with (k,a)-positive (resp., (k,a)-negative) oscillation at least μ.

We also let $P_{k,a}(\mu;q)$ and $M_{k,a}(\mu;q)$ denote the respective generating functions. Suitably modifying the procedures of Sections 3 and 4, one is able to prove the following identities.

(5.1) $$M_{k,a}(2\mu;q) = \frac{q^{\mu\{(2k+1)2\mu + (2k-2a+1)\}}}{(q)_\infty},$$

(5.2) $$M_{k,a}(2\mu-1;q) = \frac{q^{(2\mu-1)\{(2k+1)\mu - (2k-a+1)\}}}{(q)_\infty},$$

(5.3) $$P_{k,a}(2\mu;q) = \frac{q^{\mu\{(2k+1)2\mu - (2k-2a+1)\}}}{(q)_\infty},$$

(5.4) $$P_{k,a}(2\mu-1;q) = \frac{q^{(2\mu-1)\{(2k+1)\mu-a\}}}{(q)_\infty}.$$

Theorem 2 now may be extended to the following result.

Theorem 4. Let $Q_{k,a}(n)$ denote the number of partitions π of n such that $-(a-2) \le r_i(\pi) \le 2k-a-1$ for each of the successive ranks of π. Then

(5.5) $$Q_{k,a}(n) = P_{k,a}(0;n) + \sum_{\mu=1}^{\infty} (-1)^\mu M_{k,a}(\mu;n) + \sum_{\mu=1}^{\infty} (-1)^\mu P_{k,a}(\mu;n).$$

The expression on the right-hand side sieves out all partitions that have at least one rank outside the interval $[-(a-2),2k-a-1]$.

Theorem 3 now extends to the following series identity

$$\sum_{n=0}^{\infty} Q_{k,a}(n)q^n$$

(5.6)
$$= \frac{1}{(q)_\infty}\left\{1 + \sum_{n=1}^{\infty} (-1)^n q^{\frac{1}{2}n\{(2k+1)n-(2k-2a+1)\}}(1+q^{(2k-2a+1)n})\right\}$$

$$= \frac{(q^{2k+1};q^{2k+1})_\infty (q^a;q^{2k+1})_\infty (q^{2k+1-a};q^{2k+1})_\infty}{(q)_\infty}$$

$$= \prod_{\substack{n=1 \\ n \not\equiv 0,\, \pm a \pmod{2k+1}}}^{\infty} (1-q^n)^{-1},$$

where the penultimate equation follows by Jacobi's triple product identity [Hardy and Wright; 7; p. 282].

Identity (5.6) may now be used to deduce directly the following partition theorem.

Theorem 5. Let $A_{k,a}(n)$ denote the number of partitions of n into parts not congruent to 0, a, or $-a$ modulo $2k+1$. Then $A_{k,k}(n) = Q_{k,a}(n)$.

The following theorem shows how Theorem 5 reduces to the first Rogers-Ramanujan identity (1.2) when $k = 2$, $a = 2$.

Theorem 6. Let $B(n)$ denote the number of partitions of n in which the difference between any two parts is at least 2. Then $B(n) = Q_{2,2}(n)$.

Proof. Let us take a partition π of the type enumerated by $B(n)$, say $n = b_1 + b_2 + \cdots + b_s$, where $b_i - b_{i+1} \geq 2$. We form a graphical representation of π as follows. The i-th part of π is represented by a right angle of nodes: if $b_i = 2m+1$, then we form

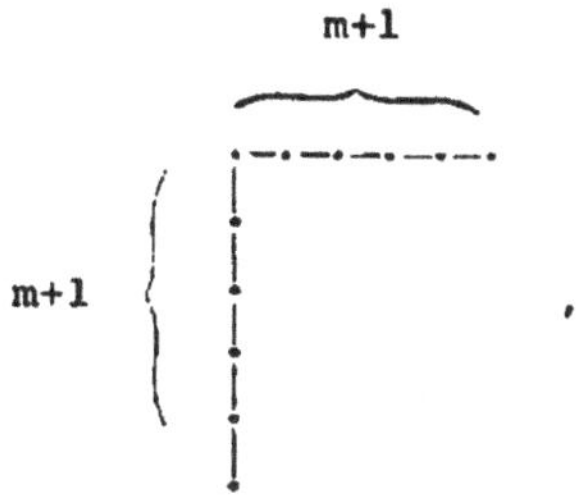

if $b_i = 2s$, then we form

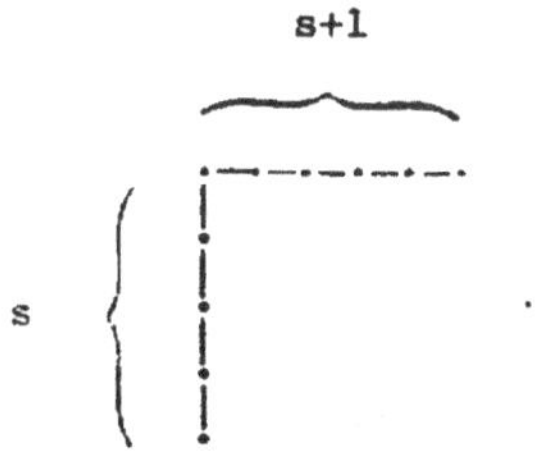

Since $b_i - b_{i+1} \geq 2$, we may collect these angles as the successive angles of a Ferrars graph in which the successive ranks are either 0 or 1, that is, a partition of the type enumerated by $Q_{2,2}(n)$. The above transformation is clearly reversible, and thus we have a one-to-one correspondence between the partitions enumerated by $B(n)$ and those enumerated by $Q_{2,2}(n)$. Hence $B(n) = Q_{2,2}(n)$. □

The following is an example of the above transformation for $n = 9$.

9 → 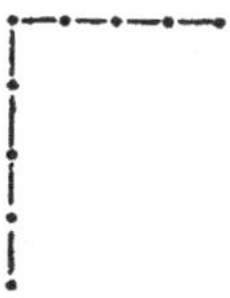 → 5 + 1 + 1 + 1 + 1

8 + 1 →  → 5 + 2 + 1 + 1

7 + 2 →  → 4 + 3 + 1 + 1

6 + 3 →  → 4 + 3 + 2

5 + 3 + 1 →  → 3 + 3 + 3.

As is well-known [Hardy-Wright; 7; p. 290-291] (and easily shown),

(5.7) $$\sum_{n=0}^{\infty} B(n)q^n = \sum_{n=0}^{\infty} \frac{q^{n^2}}{(q)_n} .$$

Hence

$$\sum_{n=0}^{\infty} \frac{q^{n^2}}{(q)_n} = \sum_{n=0}^{\infty} B(n)q^n = \sum_{n=0}^{\infty} Q_{2,2}(n)q^n$$

$$= \frac{1}{(q)_\infty}\left(1 + \sum_{n=1}^{\infty} (-1)^n q^{\frac{1}{2}n(5n-1)}(1+q^{2n})\right),$$

and this proves (1.2).

Finally the following theorem shows how Theorem 5 reduces to the second Rogers-Ramanujan identity (1.3) when $k = 2$, $a = 1$.

Theorem 7. <u>Let</u> $C(n)$ <u>denote the number of partitions of</u> n <u>in which</u> 1 <u>does not appear and the difference between any two parts is at least</u> 2. <u>Then</u> $C(n) = Q_{2,1}(n)$.

Proof. The proof here is very much like the proof of Theorem 6, so we shall omit most of the details. The only change is that if $b_1 + b_2 + \cdots + b_s$ is a partition of the type enumerated by $C(n)$ and b_i $(= 2m+1)$ is an odd part of that partition, then b_i is

represented by the angle of nodes

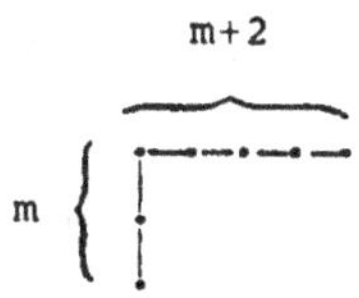

which has rank 2. The fact that $b_s > 1$ now fits in perfectly with the fact that 1 is the only positive odd number that cannot be represented by the angle of nodes of the type given above. □

We remark that the techniques we have described for the Rogers-Ramanujan identities are sufficient to prove the main theorem of [Andrews; 2], a polynomial identity that implies (1.2) and (1.3).

It is interesting to note that there is a third partition function related to Theorem 5. Namely if $B_{k,a}(n)$ denotes the number of partitions of n of the form $n = b_1 + b_2 + \cdots + b_s$ where $b_i \geq b_{i+1}$, $b_i - b_{i+k-1} \geq 2$, and $b_{s-a+1} > 1$, then B. Gordon has shown (see [Gordon; 6] and [Andrews; 1]) that

$$B_{k,a}(n) = A_{k,a}(n).$$

An independent proof that $B_{k,a}(n) = Q_{k,a}(n)$ would certainly illuminate this type of theorem.

It is hoped that the ideas presented here will stimulate research in two directions. First, partition theorems related to the successive ranks of Atkin are clearly of interest and should be studied. Second, it would be nice to find further sieves that yield other partition theorems.

REFERENCES

1. G. E. Andrews, An analytic proof of the Rogers-Ramanujan-Gordon identities, Amer. J. Math. 88 (1966), 844-846.

2. G. E. Andrews, A polynomial identity which implies the Rogers-Ramanujan identities, Scripta Math. 23 (1970), 297-305.

3. A. O. L. Atkin and H. P. F. Swinnerton-Dyer, Some properties of partitions, Proc. London Math. Soc. (3) 4 (1954), 84-106.

4. A. O. L. Atkin, A note on ranks and conjugacy of partitions, Quart. J. Math. Oxford Ser. (2) 17 (1966), 335-338.

5. F. J. Dyson, Some guesses in the theory of partitions, Eureka 8 (1944), 10-15.

6. B. Gordon, A combinatorial generalization of the Rogers-Ramanujan identities, Amer. J. Math. 83 (1961), 393-399.

7. G. H. Hardy and E. M. Wright, An Introduction to the Theory of Numbers, Fourth Edition, Oxford University Press, Oxford, 1960.

8. W. J. LeVeque, Topics in Number Theory, Vol. 1, Addison-Wesley, Reading, 1956.

9. I. J. Schur, Ein Beitag zur additiven Zahlentheorie, Sitzungsber. Akad. Wissensch., Berlin, Phys.-Math. Kl., 1917, 302-321.

10. J. J. Sylvester, A constructive theory of partitions, arranged in three acts, an interact, and an exodion, Amer. J. Math. 5 (1882), 251-330.

THE VORONOÏ SUMMATION FORMULA*

Bruce C. Berndt, University of Illinois

1. Introduction. If f is of bounded variation on $[a,b]$, then

$$\frac{1}{2}\sum_{n=a}^{b}{}' \{f(n+0) + f(n-0)\} = \int_a^b f(x)\,dx + 2\sum_{n=1}^{\infty}\int_a^b f(x)\cos(2\pi nx)\,dx ,$$

where the prime $'$ indicates that when $n = a$ only the term $\frac{1}{2}f(a+0)$ is counted and when $n = b$ only the term $\frac{1}{2}f(b-0)$ is counted. This is, of course, the famous Poisson summation formula. In 1904, Voronoï [27] formulated a conjecture which is a generalization of the Poisson formula. Let $a(n)$ be an arithmetical function. Suppose that $f(x)$ is continuous on (a,b) with only a finite number of maxima and minima there. Then there exist analytic functions $\delta(x)$ and $\alpha(x)$, depending only upon $a(n)$ and not on $f(x)$, such that

$$(1.1)\qquad \sum_{a\le n\le b}{}' a(n)f(n) = \int_a^b f(x)\delta(x)\,dx + \sum_{n=1}^{\infty} a(n)\int_a^b f(x)\alpha(nx)\,dx ,$$

where the prime $'$ has the same meaning as above. If $a(n) \equiv 1$, the conjecture is true. For then by Poisson's formula, (1.1) is valid with $\delta(x) \equiv 1$ and $\alpha(x) = 2\cos(2\pi x)$.

After a considerable effort, Voronoï established his conjecture on $0 < a < b < \infty$ for $a(n) = d(n)$, the number of positive divisors of n. In this case, (1.1) is valid with $\delta(x) = \log x + 2\gamma$, where γ denotes Euler's constant, and $\alpha(x) = 4K_0(4\pi x^{\frac{1}{2}}) - 2\pi Y_0(4\pi x^{\frac{1}{2}})$, where K_0 denotes the modified Bessel function of order zero of the second kind and Y_0 the Bessel function of order zero of the second kind. In 1929, Koshliakov [21] gave a much shorter proof of Voronoï's result, but with the assumption that f is analytic. In 1931, Dixon and Ferrar [10] gave a very simple proof under the hypothesis that $f \in C^{(2)}[a,b]$. Wilton [29] established the Voronoï formula for functions of bounded variation on $[a,b]$ and also extended the formula to include the cases $a = 0$ and $b = \infty$. A refinement of Wilton's result in the case $a = 0$ was made by Dixon and Ferrar [12].

---

*This research was partially supported by NSF Grant GP-21335.

Soon after Voronoï established his conjecture for $d(n)$, he announced [28] a corresponding result for $r(n)$, the number of representations of n as the sum of two squares. Now, in (1.1), $0 < a < b < \infty$, $\delta(x) = \pi$ and $\alpha(x) = \pi J_0(2\pi x^{\frac{1}{2}})$, where J_0 denotes the Bessel function of the first kind of order zero. Voronoï's formula was also given by Sierpinski [25]. Landau [22] established the result when f is of bounded variation on $[a,b]$, $0 \le a < b < \infty$. An extension to $b = \infty$ was made by Dixon and Ferrar [11].

Voronoï's conjecture has been subsequently established for a number of other arithmetical functions. A survey of such results in the literature yields several observations. In the derivation of the Voronoï summation formula, often an expression for $\sum_{n \le x} a(n)$ in terms of an infinite series of Bessel functions is first established. Secondly, for every Voronoï formula that has been proven, the arithmetical function $a(n)$ is generated by a Dirichlet series satisfying a functional equation involving the gamma function. A third observation is that (1.1) is usually established under generally one of three sets of hypotheses. In the first, f is "smooth", i.e., $f \in C^{(1)}[a,b]$ or $C^{(2)}[a,b]$. In the second, f is of bounded variation on $[a,b]$. In the third, f must satisfy conditions so that the theory of certain transforms is applicable. This approach was inaugurated by Ferrar [13], [14] and Guinand [16], [17], [18].

Although general theorems have been established for functions satisfying conditions that are needed for transform techniques, very few general theorems have been proven for "smooth" functions or for functions of bounded variation. The first general theorem was established by Koshliakov [20] who assumed that $\{a(n)\}$ is generated by an exponential series satisfying a "theta relation", and that f is analytic. A second general theorem is given by us in [3], where it is assumed that f is of bounded variation and $\{a(n)\}$ is generated by a Dirichlet series satisfying certain regularity conditions and a functional equation involving $\Gamma(s)$.

Our objective here is to give Voronoï summation formulae for a class of arithmetical functions generated by Dirichlet series satisfying a functional equation involving $\Gamma(s)$. One set of theorems will be stated for "smooth" functions; the other set will be stated for functions of bounded variation. We have proved similar theorems for other arithmetical functions generated by Dirichlet series satisfying a functional equation with other gamma factors [5], but for simplicity we shall confine our attention to the simplest case of $\Gamma(s)$. We shall conclude the paper with some applications to primitive characters. In particular, we shall establish some formulas for character power sums. These will enable us to evaluate the Dirichlet L-function $L(s,\chi)$ for certain positive integers.

In the sequel, s always denotes a complex number with $\sigma = \mathrm{Re}(s)$. We shall always write $\sum$ for $\sum_{n=1}^{\infty}$.

2. The arithmetical functions. Let $\{\lambda_n\}$ and $\{\mu_n\}$ denote two sequences of positive numbers strictly increasing to ∞, and $\{a(n)\}$ and $\{b(n)\}$ two sequences of complex numbers not identically zero. Suppose that the Dirichlet series

$$\varphi(s) = \sum a(n)\lambda_n^{-s} \quad \text{and} \quad \psi(s) = \sum b(n)\mu_n^{-s}$$

each converge in some half-plane and have abscissae of absolute convergence σ_a and σ_a^*, respectively. Let r be a real number. Then we say that φ and ψ satisfy the functional equation

$$\Gamma(s)\varphi(s) = \Gamma(r-s)\psi(r-s)$$

if there exists a meromorphic function $F(s)$ with the following properties:

(i) $F(s) = \Gamma(s)\varphi(s), \quad \sigma > \sigma_a,$
$F(s) = \Gamma(r-s)\psi(r-s), \quad \sigma < r - \sigma_a^*,$

(ii) $\lim_{|\mathrm{Im}(s)|\to\infty} F(s) = 0,$ uniformly in every vertical strip
$-\infty < \sigma_1 \le \sigma \le \sigma_2 < \infty;$

(iii) the poles of F are confined to some compact set.

For the arithmetical functions described above, Chandrasekharan and Narasimhan [7] have established a fundamental identity involving Bessel functions. See also our paper [4]. For $x > 0$ define

$$A(x) = \sum_{\lambda_n \le x}{}' a(n),$$

where the prime $'$ indicates that if $\lambda_n = x$, $a(n)$ is to be multiplied by $\frac{1}{2}$. Secondly, if q is a non-negative integer, define

$$Q_q(x) = \frac{1}{2\pi i} \int_{C_q} \frac{\Gamma(s)\varphi(s)}{\Gamma(s+q+1)} x^{s+q}\, ds ,$$

where C_q is a cycle enclosing all of the integrand's poles. Put

$$D(x) = A(x) - Q_0(x) .$$

Suppose that for $\sigma > \sigma_a^*$,

$$\sup_{0 \le h \le 1} \left| \sum_{m^2 \le \mu_n \le (m+h)^2} b(n)\mu_n^{\frac{1}{2}-\sigma} \right| = o(1) ,$$

as m tends to ∞. Then, if $2\sigma_a^* < r + 3/2$,

$$D(x) = \sum \frac{b(n)}{\mu_n^r} I_0(\mu_n x), \tag{2.1}$$

where for $q = 0$ or -1 ,

$$I_q(x) = x^{\frac{1}{2}(r+q)} J_{r+q}(2x^{\frac{1}{2}}) .$$

The series of Bessel functions in (2.1) converges boundedly on any compact interval in $(0,\infty)$.

3. <u>Voronoï summation formulae for "smooth" functions</u>. The main ingredients in the proofs of the theorems in this section are partial summation and integration by parts. Complete proofs may be found in our work [5].

<u>Theorem</u> 1. <u>Let</u> $f \in C^{(1)}(0,\infty)$. <u>Then, if</u> $0 < a < \lambda_1 < x < \infty$,

$$\sum_{\lambda_n \le x}{}' a(n) f(\lambda_n) = \int_a^x Q_0'(t) f(t)\,dt + \sum \frac{b(n)}{\mu_n^{r-1}} \int_a^x I_{-1}(\mu_n t) f(t)\,dt . \tag{3.1}$$

Additional hypotheses are needed to extend Theorem 1 to $a = 0$.

Theorem 2. *Let* $f \in C^{(1)}(0,\infty)$. *Suppose that* $\lim_{t\to 0} D(t)f(t) < \infty$ *and*

$$\int_0^x Q_0'(t)f(t)\,dt < \infty .$$

Furthermore, assume that $D(t)$ *converges boundedly on* $0 \le t \le x$. *Then (3.1) is valid with* $a = 0$.

The next two theorems do not depend upon an assumption that $D_0(t)$ is boundedly convergent on $0 \le t \le x$.

Theorem 3. *Let* $f \in C^{(1)}[0,\infty)$. *Suppose that the poles of* $\varphi(s)$ *lie in the half-plane* $\sigma > 0$. *Then (3.1) is valid with* $a = 0$.

Note that the hypotheses on f and φ insure the convergence of

$$\int_0^x Q_0'(t)f(t)\,dt .$$

In some applications the above condition on f at the origin is too strong.

Theorem 4. *Let* $f \in C^{(2)}(0,\infty)$. *Suppose that the poles of* $\varphi(s)$ *are in the half-plane* $\sigma > 0$. *Assume that* $Q_0(t)f(t)$ *and* $Q_1(t)f'(t)$ *are continuous at* $t = 0$. *Lastly, suppose that*

$$\int_0^1 t^{r/2 + 1/4}|f''(t)|\,dt < \infty .$$

Then (3.1) is valid with $a = 0$.

Of course, by subtraction a Voronoï formula may be given over any finite interval $[a,x]$, $a \ge 0$. We now wish to extend the above results to include the case $x = \infty$. We shall give a theorem for $a > 0$. By assuming the hypotheses of either Theorems 2, 3, or 4, we may extend the result below to include the case $a = 0$.

Theorem 5. *Let* $f \in C^{(2)}(0,\infty)$. *Suppose that* $D(x)f(x)$ *and* $x^{r/2 + 1/4} f'(x)$ *tend to* 0 *as* x *tends to* ∞. *Assume that for* $x > 0$,

$$\int_x^{\infty} t^{r/2 + 1/4} |f''(t)| dt < \infty .$$

Then, if $0 < a < \lambda_1$,

$$\lim_{x\to\infty} \left\{ \sum_{\lambda_n \le x}{}' a(n) f(\lambda_n) - \int_a^x Q_0'(t) f(t) dt \right\}$$

$$= \sum \frac{b(n)}{\mu_n^{r-1}} \int_a^{\infty} I_{-1}(\mu_n t) f(t) dt ,$$

provided that the limit on the left side exists.

To determine if $D(x)f(x)$ tends to 0 as x tends to ∞, we may consult a general theorem of Chandrasekharan and Narasimhan [8, Theorem 4.1] .

4. *Voronoï summation formulae for functions of bounded variation.* For complete proofs of the theorems in this section, see our paper [5].

Theorem 6. *Let* f *be of bounded variation on* $[a,x]$, *where* $0 < a < \lambda_1 < x < \infty$. *Then*

(4.1) $$\frac{1}{2} \sum_{\lambda_n \le x}{}' a(n) \{f(\lambda_n+0) + f(\lambda_n-0)\}$$

$$= \int_a^x Q_0'(t) f(t) dt + \sum \frac{b(n)}{\mu_n^{r-1}} \int_a^x I_{-1}(\mu_n t) f(t) dt .$$

The proof of Theorem 6 is similar to Landau's proof for the special case $a(n) = r(n)$ [22, Satz 559]. Extending Theorem 6 to include the case $a = 0$ requires some difficulty.

Theorem 7. *Let* f *be of bounded variation on* $[0,x]$. *Suppose that* $r \ge 0$ *and that the poles of both* $\varphi(x)$ *and* $\psi(s)$ *lie in the half-plane* $\sigma > 0$. *Assume that as* x *tends to* ∞,

$$\sum_{\mu_n \le x}' b(n) - \int_{C_0} \frac{\psi(s)}{s} x^s ds = O(x^{r/2}) ,$$

where C_0 *is a cycle encircling the integrand's poles. Then (4.1) is valid with* $a = 0$.

The proof of Theorem 7 depends upon showing that the series of Bessel functions in (2.1) converges boundedly on every interval $[0,x]$, $x > 0$. In order to demonstrate this, an identity for a partial sum of the infinite series in (2.1) must be established. This identity is a generalization of a theorem of Landau [22, Satz 523] and of Theorem 2 in our paper [3], but in this more general case was established by us in a slightly different way [5, Theorem 7].

We now extend Theorem 6 to include the case $x = \infty$.

Theorem 8. *Let* f *be of bounded variation on* $[a,x]$ *for every* x, $0 < a < \lambda_1 < x < \infty$, *and suppose that* f *is the integral of* f'. *Assume that* $D(x)f(x)$ *tends to* 0 *as* x *tends to* ∞,

$$\int_x^\infty t^{\sigma_a^* + \varepsilon - \frac{1}{2}} |f'(t)| dt < \infty$$

for some $\varepsilon > 0$, *and*

$$\int_x^\infty D(t) f'(t) dt < \infty .$$

Then,

$$\lim_{x\to\infty} \left\{ \frac{1}{2} \sum_{\lambda_n \le x}' a(n)\{f(\lambda_n+0) + f(\lambda_n-0)\} - \int_a^x Q_0'(t) f(t) dt \right\}$$

$$= \sum \frac{b(n)}{\mu_n^{r-1}} \int_a^\infty I_{-1}(\mu_n t) f(t) dt ,$$

provided that the series on the right side converges.

5. Additional remarks on the theorems. We indicate a few additional arithmetical functions that are covered by our theorems and are discussed in the literature. Let $\{\lambda_n\}$ denote the values, arranged in increasing order, assumed by a positive definite quadratic form in two variables. Let $a(n)$ be the number of times the form assumes the value λ_n. Voronoï was the first to give a summation formula in this case [28]. The result is also a special case of theorems of Koshliakov [20] and us [3]. Let $\{a(n)\}$ denote the coefficients of a modular cusp form of negative dimension. Then a Voronoï formula is given by Guinand [19]. Results are also special cases of general theorems of Koshliakov [20], Ferrar [14], and us [3]. Let $a(n) = F(n)$, the ideal function for an imaginary quadratic field. Then Voronoï theorems have been proved by Chandrasekharan and Narasimhan [9], Koshliakov [20], and us [3].

It would be desirable to derive Voronoï summation formulae for arithmetical functions satisfying all of the conditions of Section 2 except with the condition $2\sigma_a^* < r + 3/2$ dropped. For $q > 2\sigma_a^* - r - 3/2$, the Riesz sum

$$\sum_{\lambda_n \le x} a(n)(x-\lambda_n)^q \tag{5.1}$$

can be developed in a series of Bessel functions. See Chandrasekharan and Narasimhan's paper [7]. One might hope then to find a general formula for

$$\sum_{\lambda_n \le x} a(n) f(\lambda_n)(x-\lambda_n)^q$$

for suitable f and suitable $q > 0$, but no such formula appears to have been discovered. Sklar [26] and Senechalle [24] have actually found a formula for

$$\sum_{\lambda_n \le x} a(n) f(\lambda_n)$$

in this more general case, but their formula is not of the Voronoï type and involves the Riesz sums of (5.1).

6. Applications. The results in Sections 3 and 4 have been used to establish several arithmetical identities, many of which are useful in certain investigations in number theory. See our work [5] for many of these identities. We shall indicate some different applications here.

First, we indicate how to obtain the Poisson summation formula. It is well known [2, p. 369] that the Riemann zeta-function $\zeta(s)$ is analytic everywhere except for a simple pole at $s = 1$ with residue 1 and satisfies the functional equation

$$\pi^{-s}\Gamma(s)\zeta(2s) = \pi^{s-\frac{1}{2}}\Gamma(\tfrac{1}{2}-s)\zeta(1-2s) .$$

For simplicity, assume that the hypotheses of Theorem 1 are satisfied. Then, since $r = \frac{1}{2}$, $\lambda_n = \mu_n = \pi n^2$, $Q_0'(x) = \frac{1}{2}(\pi x)^{-\frac{1}{2}}$ and $J_{-\frac{1}{2}}(x) = (2/\pi x)^{\frac{1}{2}}\cos x$, we have

$$\sum_{\pi n^2 \le x}' f(\pi n^2) = \tfrac{1}{2}\int_a^x (\pi t)^{-\frac{1}{2}} f(t)\,dt + \sum \int_a^x (\pi t)^{-\frac{1}{2}} f(t)\cos(2n\{\pi t\}^{\frac{1}{2}})\,dt .$$

Replace a by πa^2, x by πx^2, t by πt^2, and $f(\pi x^2)$ by $f(x)$. We then obtain the Poisson formula,

$$\sum_{n \le x}' f(n) = \int_a^x f(t)\,dt + 2\sum \int_a^x f(t)\cos(2\pi n t)\,dt .$$

We shall now give some applications to characters. Throughout the sequel, $\chi(n)$ denotes a primitive, nonprincipal character modulo k. We shall let $G(m,\chi)$ denote the Gaussian sum

$$G(m,\chi) = \sum_{n=1}^{k-1} \chi(n) e^{2\pi i m n/k} ,$$

and put $G(\chi) = G(1,\chi)$.

The Dirichlet L-functions are defined for $\sigma > 0$ by

$$L(s,\chi) = \sum \chi(n) n^{-s} .$$

Define

$$b = \begin{cases} 0, & \text{if } \chi \text{ is even,} \\ 1, & \text{if } \chi \text{ is odd,} \end{cases}$$

and

$$\xi(s,\chi) = (\pi/k)^{-(s+b)/2}\Gamma(\tfrac{1}{2}\{s+b\})L(s,\chi) .$$

Then $\xi(s,\chi)$ has an analytic continuation to the entire complex plane and is an entire function of s. Furthermore [2, p. 371],

(6.1) $$\xi(s,\chi) = \varepsilon(\chi)\xi(1-s,\bar{\chi}) ,$$

where

(6.2) $$\varepsilon(\chi) = \begin{cases} k^{-\frac{1}{2}}G(\chi), & b = 0, \\ -ik^{-\frac{1}{2}}G(\chi), & b = 1. \end{cases}$$

If χ is even, in the notation of Section 2 we have $\lambda_n = \mu_n = \pi n^2/k$, $a(n) = \chi(n)$, $b(n) = \varepsilon(\chi)\bar{\chi}(n)$, and $r = \frac{1}{2}$. If χ is odd, replace s by $2s-1$ in (6.1). We then see that in the notation of Section 2, $\lambda_n = \mu_n = \pi n^2/k$, $a(n) = n\chi(n)$, $b(n) = \varepsilon(\chi)n\bar{\chi}(n)$, and $r = 3/2$.

Put $a = 0$ in (3.1). First, if χ is even, (3.1) yields

$$\sum_{\pi n^2/k \le x}{}' \chi(n)f(\pi n^2/k) = \varepsilon(\chi)\sum \bar{\chi}(n)\int_0^x (\pi t)^{-\frac{1}{2}}f(t)\cos(2n\{\pi t/k\}^{\frac{1}{2}})\,dt ,$$

since $J_{-\frac{1}{2}}(x) = (2/\pi x)^{\frac{1}{2}}\cos x$. If we replace x by $\pi x^2/k$ and let $t = u^2$, the above becomes

(6.3) $$\sum_{n \le x}{}' \chi(n)f(\pi n^2/k)$$

$$= 2\varepsilon(\chi)\pi^{-\frac{1}{2}}\sum \bar{\chi}(n)\int_0^{x(\pi/k)^{\frac{1}{2}}} f(u^2)\cos(2nu\{\pi/k\}^{\frac{1}{2}})\,du .$$

If χ is odd, (3.1) yields

$$\sum_{\pi m^2/k \le x}' n\chi(n) f(\pi m^2/k) = \varepsilon(\chi) k^{\frac{1}{2}} \pi^{-1} \sum \overline{\chi}(n) \int_0^x f(t) \sin(2n\{\pi t/k\}^{\frac{1}{2}}) dt ,$$

since $J_{\frac{1}{2}}(x) = (2/\pi x)^{\frac{1}{2}} \sin x$. If we replace x by $\pi x^2/k$ and t by u^2, the above becomes

$$(6.4) \quad \sum_{n \le x}' n\chi(n) f(\pi m^2/k)$$

$$= 2\varepsilon(\chi) k^{\frac{1}{2}} \pi^{-1} \sum \overline{\chi}(n) \int_0^{x(\pi/k)^{\frac{1}{2}}} u f(u^2) \sin(2nu\{\pi/k\}^{\frac{1}{2}}) du.$$

As a simple first example, in (6.3) let $x = k$ and $f(t) = \exp(2im\{\pi t/k\}^{\frac{1}{2}})$, where m is a positive integer. It is easily checked that the hypotheses of Theorem 7 are satisfied. Using (6.2), we find that (6.3) yields

$$G(m,\chi) = \varepsilon(\chi) \pi^{-\frac{1}{2}} \overline{\chi}(m) (\pi k)^{\frac{1}{2}} = \overline{\chi}(m) G(\chi) .$$

Hence, we have shown that if χ is even, the functional equation (6.1) implies the factorization theorem for Gaussian sums.

Next, let χ be even and put $x = k$ and $f(u) = (ku/\pi)^{m/2}$ in (6.3), where m is a positive integer. If we define

$$M_m(\chi) = \sum_{n=1}^{k-1} \chi(n) n^m ,$$

we arrive at

$$(6.5) \quad M_m(\chi) = 2\varepsilon(\chi) \pi^{-\frac{1}{2}} (k/\pi)^{m/2} \sum \overline{\chi}(n) \int_0^{(\pi k)^{\frac{1}{2}}} u^m \cos(2nu\{\pi/k\}^{\frac{1}{2}}) du .$$

Now [2, p. 313],

$$(6.6) \quad |G(\chi)|^2 = k ,$$

so that from (6.2), $\varepsilon(\chi) = k^{\frac{1}{2}}/G(\overline{\chi})$. If we let $v = 2nu\{\pi/k\}^{\frac{1}{2}}$, (6.5) then becomes

$$(6.7) \qquad G(\bar{\chi})M_m(\chi) = 2(k/2\pi)^{m+1}\sum \frac{\bar{\chi}(n)}{n^{m+1}} \int_0^{2n\pi} v^m \cos v \, dv \ .$$

Now [15, p. 421],

$$(6.8) \qquad \int_0^{2n\pi} v^m \cos v\,dv = -\sum_{j=0}^{m-1} \frac{m!}{(m-j)!}(2n\pi)^{m-j} \cos(\tfrac{1}{2}\{j+1\}\pi) \ ,$$

which is easily proved by induction on m. Hence, (6.7) becomes after simplification

$$(6.9) \quad G(\bar{\chi})M_m(\chi) = -2k^{m+1}\sum_{j=0}^{m-1} \frac{m!}{(m-j)!}(2\pi)^{-j-1} \cos(\tfrac{1}{2}\{j+1\}\pi)L(j+1,\bar{\chi}) \ .$$

From (6.9) we may calculate $L(2n,\chi)$ for any positive integer n. Letting $m = 2$ and replacing χ by $\bar{\chi}$ in (6.9), we find that

$$(6.10) \qquad L(2,\chi) = \frac{\pi^2}{k^3} G(\chi)M_2(\bar{\chi}) \ .$$

If we put $m = 4$ in (6.9) and use (6.10), we arrive at

$$L(4,\chi) = \frac{\pi^4 G(\chi)}{3k^5}\{2k^2 M_2(\bar{\chi}) - M_4(\bar{\chi})\} \ .$$

Also, observe from (6.9) that $\pi^{-2n}L(2n,\chi) \in Q(e^{2\pi i/k})$.

If χ is odd, put $x = k$ and $f(u) = (ku/\pi)^{(m-1)/2}$ in (6.4). Using (6.2) and (6.6), proceed as above to find that

$$(6.11) \qquad G(\bar{\chi})M_m(\chi) = 2i(k/2\pi)^{m+1}\sum \frac{\bar{\chi}(n)}{n^{m+1}} \int_0^{2n\pi} v^m \sin v \, dv \ .$$

By an integration by parts and the use of (6.8), we obtain after a little simplification

$$\int_0^{2n\pi} v^m \sin v\,dv = -\sum_{j=0}^{m-1} \frac{m!}{(m-j)!}(2n\pi)^{m-j}\cos(\tfrac{1}{2}j\pi) \ .$$

Substitution of the above in (6.11) yields

$$(6.12)\quad G(\overline{\chi})M_m(\chi) = -2ik^{m+1}\sum_{j=0}^{m-1} \frac{m!}{(m-j)!}(2\pi)^{-j-1}\cos(\tfrac{1}{2}j\pi)L(j+1,\overline{\chi}) .$$

From (6.12) we may calculate $L(2n-1,\chi)$ for any positive integer n. Letting $m = 1$ in (6.12), we obtain

$$L(1,\chi) = \frac{i\pi}{k^2} G(\chi)M_1(\overline{\chi}) .$$

If $m = 3$, we find that

$$L(3,\chi) = \frac{2\pi^3 i}{3k^4} G(\chi)\{k^2M_1(\overline{\chi}) - M_3(\overline{\chi})\} .$$

Note that from (6.12), we easily deduce that $\pi^{-2n+1}L(2n-1,\chi) \in Q(e^{2\pi i/k})$ for every positive integer n.

Apostol [1] has shown that $M_m(\chi)$ can be expressed as a linear combination of certain character sums involving the cotangent function. (An elementary proof of Apostol's identities has been given by us [6].) Thus, the aforementioned values of $L(n,\chi)$ can be stated in terms of certain cotangent sums.

Since the time of Euler, the values of $L(n,\chi)$ for special cases have been calculated in the literature. The first general method for calculating $L(n,\chi)$ when $n \equiv b \pmod 2$ was found by Leopoldt [23]. Leopoldt expresses $L(n,\chi)$ as a multiple of a certain generalized Bernoulli number $B^n_{\overline{\chi}}$. Our results (6.9) and (6.12) are essentially equivalent to his.

Next, put $x = k$ and $f(u) = \sin^{2m}(\pi u/k)^{\frac{1}{2}}$ in (6.3), where m is a positive integer. Using (6.2) and (6.6) again, we find that

$$(6.13)\quad G(\overline{\chi})\sum_{n=1}^{k-1} \chi(n)\sin^{2m}(\pi n/k)$$

$$= 2(k/\pi)^{\frac{1}{2}}\sum \overline{\chi}(n)\int_0^{(\pi k)^{\frac{1}{2}}} \sin^{2m}(u\{\pi/k\}^{\frac{1}{2}})\ \cos(2nu\{\pi/k\}^{\frac{1}{2}})\,du$$

$$= 2(k/\pi)\sum \overline{\chi}(n)\int_0^{\pi}\sin^{2m}v\ \cos(2nv)\,dv ,$$

upon letting $v = u(\pi/k)^{\frac{1}{2}}$. Now [15, p. 373],

$$\int_0^\pi \sin^{2m} v \cos(2nv)\,dv = \begin{cases} \frac{(-1)^n \pi}{2^{2m}} \binom{2m}{m-n}, & m \geq n, \\ 0, & m < n. \end{cases}$$

(Actually, the formula in [15] is in error as the factor π has been omitted.) Substitution of the above in (6.13) yields for χ even,

$$G(\bar{\chi}) \sum_{n=1}^{k-1} \chi(n) \sin^{2m}(\pi n/k) = \frac{k}{2^{2m-1}} \sum_{n=1}^{m} \bar{\chi}(n)(-1)^n \binom{2m}{m-n}.$$

Lastly, put $x = k$ and $f(u) = \cos^{2m}(\pi u/k)^{\frac{1}{2}}$ in (6.3), where m is a positive integer. Using the formula [15, p. 374],

$$\int_0^\pi \cos^{2m} v \cos(2nv)\,dv = \begin{cases} \frac{\pi}{2^{2m}} \binom{2m}{m-n}, & m \geq n, \\ 0, & m < n, \end{cases}$$

and proceeding as above, we find that for χ even,

$$G(\bar{\chi}) \sum_{n=1}^{k-1} \chi(n) \cos^{2m}(\pi n/k) = \frac{k}{2^{2m-1}} \sum_{n=1}^{m} \bar{\chi}(n) \binom{2m}{m-n}.$$

REFERENCES

1. T.M. Apostol, Dirichlet L-functions and character power sums, J. Number Theory 2 (1970), 223-234.

2. R. Ayoub, An Introduction to the Analytic Theory of Numbers, American Mathematical Society, Providence, 1963.

3. B.C. Berndt, Arithmetical identities and Hecke's functional equation, Proc. Edinburgh Math. Soc. 16 (1969), 221-226.

4. B.C. Berndt, Identities involving the coefficients of a class of Dirichlet series. I, Trans. Amer. Math. Soc. 137 (1969), 361-374.

5. B.C. Berndt, Identities involving the coefficients of a class of Dirichlet series. V, Trans. Amer. Math. Soc. (to appear).

6. B.C. Berndt, An elementary proof of some character sum identities of Apostol, (to appear).

7. K. Chandrasekharan and R. Narasimhan, Hecke's functional equation and arithmetical identities, Ann. of Math. 74 (1961), 1-23.

8. K. Chandrasekharan and R. Narasimhan, Functional equations with multiple gamma factors and the average order of arithmetical functions, Ann. of Math. 76 (1962), 93-136.

9. K. Chandrasekharan and R. Narasimhan, An approximate reciprocity formula for some exponential sums, Comment. Math. Helv. 43 (1968), 296-310.

10. A.L. Dixon and W.L. Ferrar, Lattice-point summation formulae, Quart. J. Math. Oxford Ser. 2 (1931), 31-54.

11. A.L. Dixon and W.L. Ferrar, Some summations over the lattice points of a circle (I), Quart. J. Math. Oxford Ser. 5 (1934), 48-63.

12. A.L. Dixon and W.L. Ferrar, On the summation formulae of Voronoï and Poisson, Quart. J. Math. Oxford Ser. 8 (1937), 66-74.

13. W.L. Ferrar, Summation formulae and their relation to Dirichlet's series, Compositio Math. 1 (1935), 344-360.

14. W.L. Ferrar, Summation formulae and their relation to Dirichlet's series II, Compositio Math. 4 (1937), 394-405.

15. I.S. Gradšteĭn and I.M. Ryžik, Table of Integrals, Series and Products, 4th ed., Fizmatgiz, Moscow, 1963; English transl., Academic Press, New York, 1965.

16. A.P. Guinand, A class of self-reciprocal functions connected with summation formulae, Proc. London Math. Soc. (2) 43 (1937), 439-448.

17. A.P. Guinand, Summation formulae and self-reciprocal functions, Quart. J. Math. Oxford Ser. 9 (1938), 53-67.

18. A.P. Guinand, Finite summation formulae, Quart. J. Math. Oxford Ser. 10 (1939), 38-44.

19. A.P. Guinand, Integral modular forms and summation formulae, Proc. Cambridge Philos. Soc. 43 (1947), 127-129.

20. N.S. Koshliakov, Application of the theory of sum-formulae to the investigation of a class of one-valued analytical functions in the theory of numbers, Messenger of Math. 58 (1929), 1-23.

21. N.S. Koshliakov, On Voronoï's sum-formula, Messenger of Math. 58 (1929), 30-32.

22. E. Landau, Vorlesungen über Zahlentheorie, Zweiter Band, Chelsea, New York, 1947.

23. H. Leopoldt, Eine Verallgemeinerung der Bernoullischen Zahlen, Abh. Math. Sem. Univ. Hamburg 22 (1958), 131-140.

24. M. Senechalle, A summation formula and an identity for a class of Dirichlet series, Acta Arith. 11 (1966), 443-449.

25. W. Sierpinski, O pewnem zagadnieniu z rachunku funkcyj asymptotycznych, Prace. Math. - Fiz. 17 (1906), 77-118.

26. A. Sklar, On some exact formulae in analytic number theory, Report of the Institute in the Theory of Numbers, University of Colorado, Boulder, Colorado, 1959, 104-110.

27. M.G. Voronoï, Sur une fonction transcendante et ses applications a la sommation de quelques séries, Ann. de l'École Norm. Sup. (3) 21 (1904), 207-267, 459-533.

28. M.G. Voronoï, Sur la développement, à l'aide des fonctions cylindriques, des sommes doubles $\Sigma f(pm^2+2qmn+rn^2)$, où $pm^2+2qmn+rn^2$ est une forme positive à coefficients entiers, Verhandlungen des Dritten Internat. Math. Kong. in Heidelberg, B.G. Teubner, Leipzig, 1905, 241-245.

29. J.R. Wilton, Voronoï's summation formula, Quart. J. Math. Oxford Ser. 3 (1932), 26-32.

THE ARITHMETIC FUNCTIONS OF THE JACOBI-PERRON ALGORITHM

Leon Bernstein, Illinois Institute of Technology

1. Introduction. Throughout this paper we shall operate in the real Euclidean space E_{n-1}, $n \geq 2$. Since we shall investigate only countable sets of vectors, we shall use the notation

$$(1.1) \qquad a^{(k)} = (a_1^{(k)}, \ldots, a_{n-1}^{(k)}) \qquad (k = 0, 1, \ldots).$$

We shall call $a^{(k)}$ a rational vector if $a_i^{(k)} \in Q$, an integral vector if $a_i^{(k)} \in J$ $(i = 1, \ldots, n-1;\ k = 0, 1, \ldots)$, an algebraic vector if $a_i^{(k)} = Q(w)$, w an algebraic irrational. Q stands for the field of the rational numbers. A transformation from E_{n-1} into E_{n-1} will be called a T-transformation if there exists a vector function in E_{n-1}, called the function associated with T, or simply a T-function,

$$(1.2) \qquad f(a^{(k)}) = b^{(k)} = (b_1^{(k)}, \ldots, b_{n-1}^{(k)}) \qquad (k = 0, 1, \ldots),$$

such that

$$(1.3) \qquad a^{(k)}T = (a_1^{(k)}-b_1^{(k)})^{-1}(a_2^{(k)}-b_2^{(k)}, \ldots, a_{n-1}^{(k)}-b_{n-1}^{(k)}, 1) = a^{(k+1)},$$
$$a_1^{(k)} \neq b_1^{(k)}, \qquad (k = 0, 1, \ldots),$$

where $a^{(0)}$ is a fixed vector in E_{n-1}. For successive transformations by T we shall use the customary notation

$$(1.4) \qquad a^{(k)}T^v = a^{(k+v)}, \quad (a^{(k)}T^v)T^u = a^{(k)}T^{u+v} = a^{(k+u+v)}$$

($k, u, v = 0, 1, \ldots;$ T^0 = identity transform). For a fixed vector $a^{(0)} \in E_{n-1}$, a given T-transformation and a given T-function we shall call the sequence $\langle a^{(0)}T^k \rangle$ the Generalized Jacobi-Perron Algorithm of $a^{(0)}$, in short GJPA, in honor of C. G. J. Jacobi [12] and Oskar Perron [14] who first formulated and investigated it. The algorithm was generalized by the author [1].

In order to state the transformation matrices, we introduce the identity matrix

$$(1.5) \qquad A^{(0)} = (\delta_{ij} = f^{(i)}(j))$$

(i, j = 0, ..., n-1; δ_{ij} is the Kronecker symbol), the transformation-coefficients

$$(1.6) \qquad f^{(i)}(n+v) = \sum_{j=0}^{n-1} b_j^{(v)} f^{(i)}(v+j)$$

(i = 0, ..., n-1; v = 0, 1, ...), with $b_0^{(v)} = a_0^{(v)} = 1$ (v = 0, 1, ...) throughout, and obtain

$$(1.7) \quad A^{(v)} = \begin{pmatrix} f^{(0)}(v) & \ldots\ldots & f^{(0)}(v+n-1) \\ \cdot & \cdot\ \cdot\ \cdot\ \cdot & \cdot \\ f^{(n-1)}(v) & \ldots\ldots & f^{(n-1)}(v+n-1) \end{pmatrix} \qquad (v = 0, 1, \ldots).$$

The following formulas were proved by the author [8]

$$(1.8) \qquad \det A^{(v)} = (-1)^{v(n-1)} \qquad (v = 0, 1, \ldots),$$

$$(1.9) \qquad a_i^{(0)} = \frac{\sum_{j=0}^{n-1} a_j^{(v)} f^{(i)}(v+j)}{\sum_{j=0}^{n-1} a_j^{(v)} f^{(0)}(v+j)} \qquad (v = 0, 1, \ldots),$$

$$(1.10) \qquad \prod_{j=1}^{v} a_{n-1}^{(j)} = \sum_{j=0}^{n-1} a_j^{(v)} f^{(0)}(v+j) \qquad (v = 0, 1, \ldots),$$

$$(1.11)\quad \det\begin{pmatrix} 1 & f^{(0)}(v+1) & \ldots & f^{(0)}(v+n-1) \\ a_1^{(0)} & f^{(1)}(v+1) & \ldots & f^{(1)}(v+n-1) \\ \vdots & \vdots & & \\ a_{n-1}^{(0)} & f^{(n-1)}(v+1) & \ldots & f^{(n-1)}(v+n-1) \end{pmatrix}$$

$$= (-1)^{v(n-1)}\Big(\sum_{j=0}^{n-1} a_j^{(0)} f^{(0)}(v+j)\Big)^{-1} \qquad (v = 0,\ 1,\ \ldots).$$

The function

$$(1.12)\qquad f^{(0)}(v) \equiv f(v) \qquad (v = 0,\ 1,\ \ldots)$$

is called the arithmetic function of GJPA, and the functions $f^{(i)}(v)$ its conjugates. It will be shown in the sequel that the conjugates are, in many cases, linear combinations of the arithmetic functions of the GJPA -- hence its predominant place in the theory of the algorithm.

The GJPA $\langle a^{(0)}T^k\rangle$ is called periodic, if there exist non-negative integers $\ell \geq 0$, $m \geq 1$ such that

$$(1.13)\qquad T^{v+m} = T^v \qquad (v = \ell,\ \ell+1,\ \ldots).$$

The sequence of vectors $b^{(0)}, \ldots, b^{(\ell-1)}$ is called the preperiod of $\langle a^{(0)}T^k\rangle$, the sequence $b^{(\ell)}, \ldots, b^{(\ell+m-1)}$ its period. If $\ell = 0$, $\langle a^{(0)}T^k\rangle$ is called purely periodic; ℓ and m are called respectively the length of the preperiod and of the period.

A T-function is called the outer T-function, if it is an integral vector, in symbols

$$(1.14)\quad f(a^{(k)}) \equiv \left[a^{(k)}\right] = \left(\left[a_1^{(k)}\right], \ldots, \left[a_{n-1}^{(k)}\right]\right) \qquad (k = 0,\ 1,\ \ldots);$$

the inner T-function, if the argument $a^{(k)}$ is an algebraic vector, and $b^{(k)}$ is a rational vector, in symbols

$$(1.15)\quad \begin{cases} f(a^{(k)}) \equiv a^{(k)}([w]'), \quad a^{(k)} = (a_1^{(k)}(w), \ldots, a_{n-1}^{(k)}(w)) \\ \qquad\qquad\qquad (w \text{ an algebraic irrational}), \\ [w]' = [w] \text{ if } w \geq 0, \quad [w]' = [w] + 1 \text{ if } w < 1. \end{cases}$$

2. <u>Previous results of the author</u>. We shall need, in the sequel, the following theorems stated by the author in previous publications [1] - [10].

<u>Theorem</u> 1. Let $F_n(x)$ be an n-th degree polynomial in x,

$$F_n(x) = x^n + k_1 x^{n-1} + \cdots + k_{n-1}x - d$$

with coefficients such that

$$(2.1)\quad \begin{cases} k_i \geq 0 \ (i = 1, \ldots, n-2), \quad d \geq 1, \quad d | k_i \quad (i = 1, \ldots, n-1), \\ k_{n-1} > cd(n+k_1+\cdots+k_{n-2}), \quad c \geq 1, \quad c \in \mathbb{R}. \end{cases}$$

Then

(i) $F_n(x)$ has a unique real root w in the positive unit interval.

(ii) $F_n(x)$ is irreducible over Q.

(iii) If

$$(2.2)\quad a^{(0)} = (w+k_1, \ w^2+k_1w+k_2, \ \ldots, \ w^{n-1}+k_1w^{n-2}+\cdots+k_{n-2}w+k_{n-1}),$$

then $(a^{(0)}T^k)$ with the outer T-function is purely periodic with length of period $m = n$ if $d > 1$, and $m = 1$ if $d = 1$. The period has the form (for $d > 1$)

$$(2.3)\quad \begin{cases} b^{(0)} = (k_1, \ldots, k_{n-1}), \quad b^{(i)} = (k_1, \ldots, k_{n-i-1}, k'_{n-i}, \ldots, k'_{n-1}), \\ k_s = dk'_s \quad (s = 1, \ldots, n-1; \quad i = 1, \ldots, n-1). \end{cases}$$

If $d = 1$, the period is of length 1 and has the form $b^{(0)} = (k_1, \ldots, k_{n-1})$.

(iv) In the algebraic field $Q(w)$ which is of degree n a unit is given by

$$\epsilon = d^{-1}w^{n} . \tag{2.4}$$

<u>Theorem</u> 2. Let $F_n(x)$ be an n-th degree polynomial in x,

$$F(x) = x^{n} + k_1 x^{n-1} + \cdots + k_{n-1}x - d,$$

with rational integral coefficients k_i such that

$$d \neq 0, \quad d|k_i \quad (i = 1, \ldots, n-1),$$

(2.5)

$$|k_{n-1}| \geq c|d|(B+2) \quad \text{where} \quad B = 1 + \sum_{i=1}^{n-2} |k_i| \geq 2, \quad c \geq (B+2)/B.$$

Then

(i) $F_n(x)$ has a unique real root w in the unit interval.

(ii) $F_n(x)$ is irreducible over Q.

(iii) If

$$a^{(0)} = (w+k_1,\ w^2+k_1w+k_2,\ \ldots,\ w^{n-1}+k_1w^{n-2}+\cdots+k_{n-1}), \tag{2.6}$$

then $(a^{(0)}T^k)$ with the inner T-function is purely periodic with length of period $m = n$ if $d \neq 1$ (d can be -1) and $m = 1$ if $d = 1$. The period has the form as in (2.3).

(iv) In the algebraic field $Q(w)$ of degree n, a unit is given by (2.4).

The reader should note that in this case, contrary to Theorem 1, the component of the vector of the period can also be negative integers; it was for this purpose of generalizing the algorithm, as originally used by Jacobi and Perron, that the inner T-function was introduced by the author. Jacobi and Perron used only the outer T-function.

Theorem 3. Let the n-th degree algebraic number field $Q(w)$ be presented by

$$w = \sqrt[n]{D^n + d}; \quad d,\ D \ \text{natural numbers}, \quad d \mid D,$$

(2.7)

$$D \geq (\ell-2)(n-1)d \quad \text{if} \quad n \leq 4, \quad D \geq (n-2)d \quad \text{if} \quad n > 4,$$

and let

$$(2.8) \quad a^{(0)} = \left(\ldots, \sum_{i=0}^{s} \binom{n-s-1+i}{i} D^i w^{s-i}, \ldots\right) \quad (s = 1, \ldots, n-1).$$

Then the $\langle a^{(0)} T^k \rangle$ with the outer T-function is purely periodic, and the period has length $m = n$ for $d > 1$, $m = 1$ for $d = 1$. The period has the form (for $d > 1$)

$$(2.9) \quad \begin{cases} b^{(0)} = \left(\binom{n}{1}D, \binom{n}{2}D^2, \ldots, \binom{n}{n-1}D^{n-1}\right), \\ b^{(i)} = \left(\binom{n}{1}D, \ldots, \binom{n}{n-i-1}D^{n-i-1}, \binom{n}{n-i}d^{-1}D^{n-i}, \ldots, \binom{n}{n-1}d^{-1}d^{n-1}\right) \end{cases}$$

$(i = 1, \ldots, n-1)$.

For $d = 1$ the form of $b^{(0)}$ is (2.9). In $Q(w)$ the following $\tau(n)$ algebraic integers are units

$$(2.10) \qquad \epsilon_s = \frac{(w - D)^s}{w^s - D^s}, \quad s \mid n, \quad s > 1.$$

Theorem 4. [Stender [15], Bernstein [3]]. In the algebraic number field $Q = (w)$,

$$(2.11) \quad w = \sqrt[3]{D^3 + k}, \quad k \ \text{a rational integer such that} \ k \mid 3D^2, \quad |k| > 1,$$

the algebraic integer

$$(2.12) \qquad \epsilon - \frac{(w - D)^3}{k}$$

is a basic unit with the only exception $D = k = 2$, where $\epsilon = \frac{(w-2)^3}{2} = 1 + 6w - 3w^2$, and $\sqrt{\epsilon} = \frac{1}{3}(-7 + w + 2w^2)$ is the basic unit.

By Dirichlet, there is only one basic unit in $Q(w)$, $w = \sqrt[3]{m}$, m a rational number, not a perfect cube. For $|k| = 1$, $w = \sqrt[3]{D^3 \pm 1}$, since $\epsilon = \pm \frac{(w-1)}{1}$, and $w - D \in Q$ is a unit. Already Nagell has proved that $w - D$ is the basic unit of $Q(w)$, with the only exception $w = \sqrt[3]{3^3 + 1} = \sqrt[3]{28}$, where $\epsilon' = \frac{1}{3}(-1 - w + \frac{1}{2}w^2)$ is the basic unit, and $\epsilon'^2 = w - 3$.

Theorem 5. [6] Let $w = \sqrt[n]{D^n + k}$, $k | D^n$ and $k | pD^n$, if $n = p^u$, $u \geq 1$, p a prime, D a natural number and k a rational integer. Then

$$\epsilon = \frac{(w - D)^n}{k} \tag{2.13}$$

is a unit in $Q(w)$.

Theorem 6. Let $w = \sqrt[n]{D^n + d}$, $d \mid D$, D, d rational integers, $D > 1$, $|d| > 0$, $m = w^n$ numbers. The norm $D(x_1, x_2, \ldots, x_n; m)$ of any algebraic numbers $\alpha = x_1 + x_2 w + \cdots + x_n w^{n-1} \in Q(\alpha)$ equals

$$D(x_1, \ldots, x_n; m) = \begin{vmatrix} x_1 & x_2 & \cdots & x_n \\ mx_n & x_1 & \cdots & x_{n-1} \\ \cdots & \cdots & \cdots & \cdots \\ mx_2 & mx_3 & \cdots \; mx_n & x_1 \end{vmatrix} . \tag{2.14}$$

Infinitely many solutions of

$$D(x_1, \ldots, x_n; m) = 1 \tag{2.15}$$

are given by the formula

(2.16) $$x_{s,t} = \sum_{i=0}^{n-s} \binom{n-s}{i} D^i f(tn+s-1+i)$$

$(s = 1, \ldots; n;\ t = 1, 2, \ldots)$ where the arithmetic function $f(v)$ is calculated from

$$\langle a^{(0)} T^k \rangle, \quad a^{(0)} = (\ldots, \sum_{i=0}^{s} \binom{n-s-1+i}{i} W^{s-i} D^i, \ldots),$$

with the inner T-function.

Note. It should be emphasized that here, in the GJPA of $a^{(0)}$, the inner T-function is being used, and that only the restriction $d|D$ holds; it should further be noted that d can be negative so that also $m = D^n - d$, $(d \geq 1)$, though in this case infinitely many solutions of (2.15) can be obtained using the outer T-function. This highly complicated case has been dealt with by the author in [3] and will be omitted here.

3. The arithmetic function of GJPA and its conjugates. In this chapter we shall express the conjugates $f^{(i)}(v)$ as linear functions of the arithmetic function $f^{(0)}(v)$. We obtain from Theorem 2, with $c \geq 2(B+2)/B$ (and taking into account that $\langle a^{(0)} T^v \rangle$ is purely periodic with length of period $= n$ for $|d| > 1$), and from formula (1.9),

(3.1) $$a_i^{(0)} = \frac{\sum_{j=0}^{n-1} a_j^{(0)} f^{(i)}(tn+j)}{\sum_{j=0}^{n-1} a_j^{(0)} f(tn+j)}$$

$(i = 0, 1, \ldots, n-1;\ t = 0, 1, \ldots;\ f^{(0)}(v) = f(v))$.

The following formulas are easily verified

$$(3.2) \qquad wa_i^{(0)} + k_{i+1} = a_{i+1}^{(0)}$$

$(i = 0, \ldots, n-2;\ a_0^{(0)} = 1;\ \ wa_{n-1}^{(0)} = d)$.

We obtain from (3.1), multiplying by w and adding k_{i+1}, and in virtue of (3.2) and again (3.1), for $i + 1$.

$$\frac{\sum_{j=0}^{n-1} a_j^{(0)} f^{(i+1)}(tn+j)}{\sum_{j=0}^{n-1} a_j^{(0)} f^{(i+1)}(tn+j)} = \frac{\sum_{j=0}^{n-1} wa_j^{(0)} f^{(i)}(tn+j)}{\sum_{j=0}^{n-1} a_j^{(0)} f(tn+j)} + k_{i+1}$$

$$(i = 0,\ 1,\ \ldots,\ n-2),$$

$$\sum_{j=0}^{n-1} a_j^{(0)} f^{(i+1)}(tn+j) = \sum_{j=0}^{n-1} (wa_j^{(0)} f^{(i)}(tn+j) + k_{i+1}a_j^{(0)} f(tn+j)).$$

But, from (3.2), $wa_j^{(0)} = a_{j+1}^{(0)} - k_{j+1}$ $(j = 0,\ 1,\ \ldots,\ n-2)$, so that

$$\sum_{j=0}^{n-1} a_j^{(0)} f^{(i+1)}(tn+j)$$

$$= \sum_{j=0}^{n-2} (a_{j+1}^{(0)} - k_{j+1}) f^{(i)}(tn+j) - wa_{n-1}^{(0)} f^{(i)}(tn+n-1)$$

$$+ k_{i+1} \sum_{j=0}^{n-1} a_j^{(0)} f(tn+j)$$

$$= \sum_{j=0}^{n-2} a_{j+1}^{(0)} f^{(i)}(tn+j) - \sum_{j=0}^{n-2} k_{j+1} f^{(i)}(tn+j) + df^{(i)}(tn+n-1)$$

$$+ k_{i+1} \sum_{j=0}^{n-1} a_j^{(0)} f(tn+j),$$

$$= \sum_{j=1}^{n-1} a_j^{(0)} f^{(i)}(tn+j-1) - \sum_{j=0}^{n-2} k_{j+1} f^{(i)}(tn+j) + df^{(i)}(tn+n-1)$$

$$+ k_{i+1} \sum_{j=0}^{n-1} a_j^{(0)} f(tn+j);$$

$$f^{(i+1)}(tn) + \sum_{j=1}^{n-1} a_j^{(0)} f^{(i+1)}(tn+j)$$

$$= \sum_{j=1}^{n-1} a_j^{(0)} f^{(i)}(tn+j-1) - \sum_{j=0}^{n-2} k_{j+1} f^{(i)}(tn+j) + df^{(i)}(tn+n-1)$$

$$+ k_{i+1} f(tn) + k_{i+1} \sum_{j=1}^{n-1} a_j^{(0)} f(tn+j).$$

The following reasoning is now decisive for our purposes: since w is a n-th degree algebraic irrational by hypothesis, so are the numbers $a_i^{(0)}$ $(i = 1, \ldots, n-1)$, being polynomials in w of degree $< n$. Therefore, any linear equation in the $a_i^{(0)}$ must vanish identically, and we obtain from the last equation:

$$(3.3)\quad \begin{cases} f^{(i+1)}(tn) = df^{(i)}(tn+n-1) \\ \qquad - \sum\limits_{j=1}^{n-1} k_j f^{(i)}(tn+j-1) + k_{i+1} f(tn); \\ f^{(i+1)}(tn+j) = f^{(i)}(tn+j-1) + k_{i+1} f(tn+j) \end{cases}$$

$$(j = 1, \ldots, n-1; \quad i = 0, 1, \ldots, n-2).$$

From (3.1) we obtain, for $i = n - 1$, multiplying by w and in virtue of (3.2),

$$d = \frac{\sum\limits_{j=0}^{n-1} wa_j^{(0)} f^{(n-1)}(tn+j)}{\sum\limits_{j=0}^{n-1} a_j^{(0)} f(tn+j)},$$

$$d \sum_{j=0}^{n-1} a_j^{(0)} f(tn+j)$$

$$= \sum_{n=0}^{n-2} (a_{j+1}^{(0)} - k_{j+1}) f^{(n-1)}(tn+j) + df^{(n-1)}(tn+n-1),$$

$$df(tn) + d \sum_{j=1}^{n-1} a_j^{(0)} f(tn+j)$$

$$= \sum_{j=1}^{n-1} (a_j^{(0)} - k_j) f^{(n-1)}(tn+j-1) + df^{(n-1)}(tn+n-1).$$

Hence, by the same reasoning as before,

$$f^{(n-1)}(tn+j-1) = df(tn+j) \quad (j = 1, \ldots, n-1),$$

$$df^{(n-1)}(tn+n-1) = df(tn) + \sum_{j=1}^{n-1} k_j f^{(n-1)}(tn+j-1)$$

$$= df(tn) + \sum_{j=1}^{n-1} k_j df(tn+j),$$

so that

$$(3.4) \quad \begin{cases} f^{(n-1)}(tn+j-1) = df(tn+j) \quad (j=1,\ldots,n-1), \\ f^{(n-1)}(tn+n-1) = f(tn) + \sum_{j=1}^{n-1} k_j f(tn+j) \\ \qquad = \sum_{j=0}^{n-1} k_j f(tn+j) = f(tn+n). \end{cases}$$

From (3.3) we obtain, for $i = 0$,

$$f^{(1)}(tn) = df(tn+n-1) - \sum_{j=1}^{n-1} k_j f(tn+j-1) + k_1 f(tn)$$

$$= df(tn+n-1) - \sum_{j=2}^{n-1} k_j f(tn+j-1),$$

so that

$$
(3.5) \quad \begin{cases} f^{(1)}(tn) = df(tn+n-1) - \sum_{j=2}^{n-1} k_j f(tn+j-1), \\ \\ f^{(1)}(tn+j) = f(tn+j-1) + k_1 f(tn+j) \\ \qquad (j = 1, \ldots, n-1; \quad t = 1, 2, \ldots). \end{cases}
$$

Formula (3.5) completely resolves the problems of expressing the first conjugate by its arithmetic function. We further obtain from (3.3), for $i = 1$,

$$
f^{(2)}(tn) = df^{(1)}(tn+n-1) - \sum_{j=1}^{n-1} k_j f^{(1)}(tn+j-1) + k_2 f(tn),
$$

$$
f^{(2)}(tn+j) = f^{(1)}(tn+j-1) + k_2 f(tn+j) \qquad (j = 1, \ldots, n-1).
$$

Hence, taking into account (3.5),

$$
\begin{aligned}
f^{(2)}(tn) &= d[f(tn+n-2) + k_1 f(tn+n-1)] \\
&\quad - k_1\Big[df(tn+n-1) - \sum_{j=2}^{n-1} k_j f(tn+j-1)\Big] \\
&\quad - \sum_{j=2}^{n-1} k_j [f(tn+j-2) + k_1 f(tn+j-1)] + k_2 f(tn) \\
&= df(tn+n-2) + \sum_{j=2}^{n-1} k_1 k_j f(tn+j-1) \\
&\quad - \sum_{j=2}^{n-1} k_j f(tn+j-2) - \sum_{j=2}^{n-1} k_1 k_j f(tn+j-2) + k_2 f(tn)
\end{aligned}
$$

$$= df(tn+n-2) - \sum_{j=2}^{n-1} k_j f(tn+j-2) + k_2 f(tn)$$

$$= df(tn+n-2) - \sum_{j=3}^{n-1} k_j f(tn+j-2),$$

$$f^{(2)}(tn+1) = f^{(1)}(tn) + k_2 f(tn+1)$$

$$= df(tn+n-1) - \sum_{j=3}^{n-1} k_j f(tn+j-1),$$

$$f^{(2)}(tn+j) = f(tn+j-2) + k_1 f(tn+j-1) + k_2 f(tn+j) \qquad (j = 2, \ldots, n-1).$$

From these results, we have

$$(3.6) \begin{cases} f^{(2)}(tn) = df(tn+n-2) - \sum_{j=3}^{n-1} k_j f(tn+j-2), \\ f^{(2)}(tn+1) = df(tn+n-1) - \sum_{j=3}^{n-1} k_j f(tn+j-1), \\ f^{(2)}(tn+j) = f(tn+j-2) + k_1 f(tn+j-1) + k_2 f(tn+j) \\ \qquad\qquad (j = 2, \ldots, n-1). \end{cases}$$

From (3.5), (3.6) we see how $f^{(1)}(tn+j)$, $f^{(2)}(tn+j)$ are expressed linearly by $f(tn+j)$, $j = 0, \ldots, n-1$, so that indeed $f^{(1)}(v)$ and $f^{(2)}(v)$ are expressed linearly by the arithmetic function for any v. Proceeding recursively in this way, the reader will now easily verify, by induction, the general formulas

$$
(3.7)\quad \begin{cases} f^{(i)}(tn+s) = df(tn+n-i+s) - \sum\limits_{j=i+1}^{n-1} k_j f(tn+j-i+s) \\ \qquad\qquad (s = 0, \ldots, i-1), \\ f^{(i)}(tn+j) = \sum\limits_{u=0}^{i} k_u f(tn+j-i+u) \\ \qquad\qquad (j = i, \ldots, n-1;\ k_o = 1;\ i = 1, \ldots, n-2). \end{cases}
$$

Comparing (3.7) with (3.4), the reader will also verify that formula (3.7) holds also for the case $i = n-1$, if we put $\sum\limits_{j=i+1}^{n-1} = \sum\limits_{j=n}^{n-1} = 0$. Thus formula (3.7) completely solves the case of expressing the conjugates $f^{(i)}(v)$ linearly by the arithmetic function $f(v)$. It should be noted that formulas (3.7) also hold, if we drop the condition that w is an n-th degree irrational.

4. <u>The basic theorem</u>. The problem of expressing the arithmetic function $f(v)$ of the GJPA explicitly by the argument v is unsolved. It is, therefore, quite suprising that a solution to this problem exists in the case of a periodic GJPA of length n (or one), as was shown by Helmut Hasse and the author in a joint paper published recently [10]. In this paper the arithmetic function and all its $n-1$ conjugates were stated explicitly as functions of the period vectors. Since, as we have shown in the previous section, the conjugates can be expressed linearly by the arithmetic function, we shall give here the solution for this function only. This is stated in

<u>Theorem</u> 7. Let a GJPA be purely periodic with length of period n having the form

$$
b^{(0)} = (k_1, \ldots, k_{n-1}), \quad b^{(i)} = (k_1, \ldots, k_{n-i-1}, k'_{n-i}, \ldots, k'_{n-1}),
$$

$$
k_s = d\,k'_s, \qquad (i = 1, \ldots, n-1;\ s = 1, \ldots, n-1).
$$

Then the arithmetic function of this GJPA is given by the formula

$$(4.1)\quad \begin{cases} f((s+1)n+j) = \\ d^s \sum\limits_{L=sn+j} \binom{x_0+x_1+\cdots+x_{n-1}}{x_0,\ldots,x_{n-1}} d^{-(x_0+x_1+\cdots+x_{n-1})} k_1^{x_1} k_2^{x_2} \cdots k_{n-1}^{x_{n-1}}, \\ L = nx_0 + (n-1)x_1 + \cdots + 2x_{n-2} + 1\cdot x_{n-1}, \end{cases}$$

where the x_i are non-negative integers $(i = 0, 1, \ldots, n-1)$; $s = 0, 1, 2, \ldots$; $j = 0, 1, \ldots, n-1$, and

$$\binom{x_0+x_1+\cdots x_{n-1}}{x_0, x_1, \ldots x_{n-1}} = \frac{(x_0+x_1+\cdots+x_{n-1})!}{x_0!x_1!\cdots x_{n-1}!}.$$

For $d = 1$, the period is of length one; formula (4.1) remains the same for the calculation of $f(v)$, taking on a much simpler form. As is seen from (4.1), the summation sign is extended over the solution of the linear Diophantine equation

$$L = nx_0 + (n-1)x_1 + \cdots + 1\cdot x_{n-1},$$

which becomes quite complicated for large L. The reader will recognize in $f((s+1)n+j)$ the partition function of L.

A few interesting formulas hold for specifying the values of the k_i. In the first case, let

$$(4.2)\quad \begin{gathered} k_1 = k_2 = \cdots = k_{n-2} = 0, \\ d^{-1}k_{n-1} = b \neq 0, \quad b^n = z, \quad zd = x. \end{gathered}$$

Then the arithmetic function of this GJPA is given by the formula

$$(4.3)\qquad f((s+1)n+j) = b^j \sum_{i=0}^{s} \binom{(s-i)n+j+i}{i} x^i$$

$(s = 0, 1, \ldots;\ j = 0, 1, \ldots, n-1)$. As in the general case, this formula is derived from the generating function

$$(4.4)\qquad \frac{1}{d - k_{n-1}x - x^n} = \sum_{v=0}^{\infty} d^{-\left[\frac{v+n}{n}\right]} f(v+n)x^v.$$

In the second case, let

$$(4.5)\qquad k_1 = k_2 = \cdots = k_{n-1} = d = 1.$$

Formula (4.1) takes the form

$$(4.6)\qquad f((s+1)n+j) = \sum_{L=sn+j} \binom{x_0+x_1+\cdots+x_{n-1}}{x_0,\ldots,x_{n-1}},$$

where $L = nx_0 + (n-1)x_1 + \cdots + x_{n-1}$, $s = 0, 1, \ldots;$ $j = 0, 1, \ldots, n-1$.

In another paper [6], the author has proved by induction the interesting formula (which, of course, can also be derived by the general generating function)

$$f((s+1)n+j)$$

$$(4.7)\qquad = 2^{j-s-1}\sum_{i=0}^{s} (-1)^i\left[\binom{(s-i)n+j}{i} + \binom{(s-i)n+j-1}{i-1}\right]2^{(s-i)(n+1)}$$

$(s = 0, 1, \ldots;\ j = 0, 1, \ldots, n-1)$. We have thus obtained the interesting identity

$$(4.8)\qquad \begin{cases} \displaystyle\sum_{L=sn+j} \binom{x_0+x_1+\cdots+x_{n-1}}{x_0,\ldots,x_{n-1}} \\ \displaystyle\qquad = 2^{n-s-1}\sum_{i=0}^{s} (-1)^i \left[\binom{(s-i)n+j}{i} + \binom{(s-i)n+j-1}{i-1}\right]2^{(s-i)(n+1)}, \end{cases}$$

$L = nx_0 + \cdots + x_{n-1}$, $s = 0, 1, \ldots;$ $j = 0, 1, \ldots, n-1$.

5. <u>About Hilbert's Tenth Problem</u>. The basic technique in the solution of Hilbert's Tenth Problem, as given by J. Matijasevič [13] and generalized by Martin Davis [11], essentially hinges on the following considerations: a basic unit in any real quadratic number field is given by

$$\epsilon = D + w, \quad w = \sqrt{D^2 - 1}, \tag{5.1}$$

where D is a natural number > 1 and we set $m = D^2 - 1$.

Let the powers of ϵ have the form

$$\begin{cases} \epsilon^n = x_n + y_n w \quad (n = 1, 2, \ldots), \\ x_0 = 1, \ y_0 = 0, \ x_1 = D, \ y_1 = 1. \end{cases} \tag{5.2}$$

A recursion formula for x_n, y_n is easily obtained from

$$\begin{aligned} x_{n+1} + y_{n+1}w &= (x_n + y_n w)(D+w) \\ &= Dx_n + my_n + (x_n + Dy_n)w, \end{aligned}$$

so that

$$\begin{aligned} x_{n+1} &= Dx_n + my_n, \\ y_{n+1} &= x_n + Dy_n \end{aligned}$$

and, as the reader can easily verify, the following general recurrence formula holds:

$$\begin{cases} u_{n+2} - 2Du_{n+1} + u_n = 0, \\ u_n = x_n \quad \text{with} \quad x_0 = 1, \ x_1 = D, \\ u_n = y_n \quad \text{with} \quad y_0 = 0, \ y_1 = 1. \end{cases} \tag{5.3}$$

This formula is used repeatedly by Matijasevic and Davis. Of course, the powers of ϵ, viz. $(w+d)^n$ can be calculated directly.

In algebraic number fields of higher degree the situation is rather complicated; explicit units are rarely known, the only exception being the fields of special structure and investigated by the author (see Theorem 1 and [8]). It has been shown previously that in the field

$$F = Q(w), \quad w = \sqrt[n]{m}, \quad m = D^n - 1, \quad n \geq 2, \quad D \text{ a natural number } > 1,$$

the algebraic integers

$$(5.4) \qquad \begin{aligned} \epsilon &= (D - w)^{-1} = D^{n-1} + D^{n-2}w + \cdots + w^{n-1}, \\ \epsilon^k &= x_{1,k} + x_{2,k}w + \cdots + x_{n,k}w^{n-1} \quad (k = 1, 2, \ldots), \end{aligned}$$

are units with positive integral coefficients. Of course, one could set out with the unit $\epsilon^{-1} = D - w$, but its powers would also have negative coefficients. We shall find a recursive formula for the coefficients of the powers of ϵ in the field of degree $n \geq 2$, analogous to that of (5.3) in the quadratic field. From

$$(5.5) \qquad \begin{aligned} &x_{1,k+1} + x_{2,k+1}w + \cdots + x_{n,k+1}w^{n-1} \\ &= (x_{1,k} + x_{2,k}w + \cdots + x_{n,k}w^{n-1})(D-w)^{-1} \end{aligned}$$

we obtain

$$(5.6) \qquad \begin{aligned} &(D - w)(x_{1,k+1} + x_{2,k+1}w + \cdots + x_{n,k+1}w^{n-1}) \\ &= x_{1,k} + x_{2,k}w + \cdots + x_{n,k}w^{n-1}, \end{aligned}$$

hence, by comparison of the coefficients of powers of w which is an n-th degree algebraic irrational,

$$(5.1)\qquad \begin{cases} Dx_{1,k+1} - mx_{n,k+1} = x_{1,k}\,, \\ Dx_{2,k+1} - x_{1,k+1} = x_{2,k}\,, \\ Dx_{3,k+1} - x_{2,k+1} = x_{3,k}\,, \\ \cdots \\ Dx_{n,k+1} - x_{n-1,k+1} = x_{n,k}. \end{cases}$$

Introducing the matrices

$$(5.8)\qquad U_k = \begin{pmatrix} x_{1,k} \\ x_{2,k} \\ \cdot \\ \cdot \\ \cdot \\ x_{n,k} \end{pmatrix}, \qquad W = \begin{pmatrix} D & 0 & . & . & . & . & 0 & m \\ -1 & D & 0 & . & . & . & 0 & 0 \\ 0 & -1 & D & 0 & . & . & 0 & 0 \\ \cdot \\ \cdot \\ \cdot \\ 0 & . & . & . & . & 0 & -1 & D \end{pmatrix},$$

we obtain from (5.7), with the notation of (5.8),

$$(5.9)\qquad U_k = WU_{k+1} \quad (k = 0,\ 1,\ \ldots).$$

Since $U_{k-1} = WU_k = W(WU_{k+1}) = W^2U_{k+1}$, we obtain, with E denoting the identity matrix,

$$(5.10)\qquad \begin{cases} U_{k+1} = EU_{k+1} \\ U_k = WU_{k+1} \\ U_{k-1} = W^2U_{k+1} \\ U_{k-2} = W^3U_{k+1} \\ \cdot \\ \cdot \\ \cdot \\ U_{k-n+1} = W^nU_{k+1} \end{cases}$$

The characteristic equation of W is given by

$$(5.11) \quad \det W(\lambda) = \begin{vmatrix} D-\lambda & 0 & & & 0 & -m \\ -1 & D-\lambda & & & 0 & 0 \\ 0 & -1 & D-\lambda & & 0 & 0 \\ \cdots & \cdots & \cdots & \cdots & \cdots & \cdots \\ 0 & & & & & \\ 0 & 0 & 0 & 0 & -1 & D-\lambda \end{vmatrix} = 0.$$

But

$$\det W(\lambda) = (D-\lambda)^n + \begin{vmatrix} 0 & 0 & . & . & . & . & 0 & -m \\ -1 & D-\lambda & 0 & . & . & . & 0 & 0 \\ 0 & -1 & D-\lambda & 0 & . & & 0 & 0 \\ \vdots & & & & & & & \\ 0 & 0 & . & . & . & 0 & -1 & D-\lambda \end{vmatrix}$$

$$= (D-\lambda)^n + (-1)^{n-2}(-m) \begin{vmatrix} -1 & D-\lambda & 0 & . & . & . & 0 \\ 0 & 1 & D-\lambda & & 0 & & 0 \\ \vdots & & & & & & \\ 0 & 0 & . & . & . & & -1 \end{vmatrix}$$

$$= (D-\lambda)^n + (-1)^{n-2}(-m)(-1)^{n-2} = (D-\lambda)^n - m.$$

We have thus obtained the characteristic equation for W,

$$(D-\lambda)^n - m = 0,$$

$$D^n - \binom{n}{1}D^{n-1}\lambda + \binom{n}{2}D^{n-2}\lambda^2 - \cdots + (-1)^n\lambda^n - (D^n-1) = 0,$$

hence

$$(-1)^n\lambda^n + (-1)^{n-1}\binom{n}{1}D\lambda^{n-1} + (-1)^{n-2}\binom{n}{2}D^2\lambda^{n-2} + \cdots + (-1)D^{n-1}\lambda + 1 = 0$$

or

$$\lambda^n - \binom{n}{1}D\lambda^{n-1} + \binom{n}{2}D^2\lambda^{n-2} - \cdots + (-1)^n = 0. \tag{5.12}$$

From (5.12) we now obtain, by the Cayley-Hamilton Theorem.

$$W^n - \binom{n}{1}DW^{n-1} + \binom{n}{2}D^2W^{n-2} - \cdots + (-1)^n E = Z, \tag{5.13}$$

and from (5.13), (5.10),

$$\begin{aligned} &U_{k-n+1} - \binom{n}{1}DU_{k-n+2} + \binom{n}{2}D^2U_{k-n+3} \\ &\qquad - \cdots + (-1)^{n-1}\binom{n}{1}D^{n-1}U_k + (-1)^n U_{k+1} = 0. \end{aligned} \tag{5.14}$$

Substituting in (5.14) $k+n-1$ for k, we obtain

$$\begin{aligned} &U_{k+n} - \binom{n}{1}D^{n-1}U_{k+n-1} + \binom{n}{2}D^{n-2}U_{k+n-2} \\ &\qquad - \cdots + (-1)^{n-1}\binom{n}{n-1}DU_{k+1} + (-1)^n U_k = 0, \end{aligned} \tag{5.15}$$

which is the analogous recurrence equation for Hilbert's Tenth Problem in any dimension $n \geq 2$. For $n = 2$ we obtain

$$U_{k+2} - 2DU_{k+1} + U_k = 0,$$

which is indeed (5.3). Of course, (5.15) is a vector equation and is equivalent to

$$\begin{cases} x_{1,k+n} - \binom{n}{1}D^{n-1}x_{1,k+n-1} + \binom{n}{2}D^{n-2}x_{1,k+n-2} \\ \qquad - \cdots + (-1)^n x_{1,k} = 0, \\ x_{2,k+n} - \binom{n}{1}D^{n-1}x_{2,k+n-1} + \binom{n}{2}D^{n-2}x_{2,k+n-2} \\ \qquad - \cdots + (-1)^n x_{2,k} = 0, \\ \cdots\cdots\cdots\cdots\cdots\cdots\cdots\cdots \\ x_{n,k+n} - \binom{n}{1}D^{n-1}x_{n,k+n-1} + \binom{n}{2}D^{n-2}x_{n,k+n-2} \\ \qquad - \cdots + (-1)^n x_{n,k} = 0. \end{cases} \tag{5.16}$$

In order to use the formulas (5.16), we still need the n initial values of $x_{i,0}, x_{i,1}, \ldots, x_{i,n}$, $(i = 1, \ldots, n)$; these are obtained from the successive powers of

$$\epsilon^t = (D - w)^{-t} = (D^{n-1} + D^{n-2}w + \cdots + w^{n-1})^t$$

$(t = 0, 1, \ldots, n-1)$. For $n = 3$, we obtain

$$\epsilon^0 = 1 + 0w + 0w^2,$$

$$\epsilon^1 = D^2 + Dw + 1w^2,$$

$$\epsilon^2 = (D^2 + Dw + w^2)^2 = D^4 + D^2w^2 + w^4 + 2D^3w + 2D^2w^2 + Dw^3$$

$$= (2D^4 - D) + (3D^3 - 1)w + 3D^2w^2,$$

hence

$$\begin{pmatrix} x_{1,0} \\ x_{1,1} \\ x_{1,2} \end{pmatrix} = \begin{pmatrix} 1 \\ D^2 \\ 2D^4 - D \end{pmatrix}, \quad \begin{pmatrix} x_{2,0} \\ x_{2,1} \\ x_{2,2} \end{pmatrix} = \begin{pmatrix} 0 \\ D \\ 3D^3 - 1 \end{pmatrix},$$

$$\begin{pmatrix} x_{3,0} \\ x_{3,1} \\ x_{3,2} \end{pmatrix} = \begin{pmatrix} 0 \\ 1 \\ 3D^2 \end{pmatrix}.$$

In [3], I proved that for

$$(5.17) \qquad \begin{cases} w = \sqrt[n]{D^n - 1}, \quad D \geq 2(n-1), \quad n \geq 2, \\ a^{(0)} = \left(\ldots, \sum_{i=0}^{s} \binom{n-s-1+i}{i} (D-1)^i w^{s-i}, \ldots\right) \end{cases}$$

$(s = 1, \ldots, n-1)$,

$\langle a^{(0)}T^k \rangle$ with the outer T-function is purely periodic with length of period n. It is further shown that, in this case

$$\prod_{i=1}^{n} a_{n-1}^{(i)} = \frac{1}{D - w} = \epsilon .$$

From (1.10) we then obtain

$$\epsilon^k = (D - w)^{-k} = \sum_{s=0}^{n-1} \left(\sum_{i=0}^{s} \binom{n-s-1+i}{i} (D-1)^i w^{s-i} \right) f(nk+s)$$

(5.18)

$$= x_{1,k} + x_{2,k} w + \cdots + x_{n,k} w^{n-1} .$$

Rearranging the polynomials under the summation sign into increasing powers of w, we obtain easily from (5.18), by comparison of coefficients,

$$(5.19) \quad x_{i,k} = \sum_{j=0}^{n-i} \binom{n-i}{j} (D-1)^j f(nk+i-1+j) \qquad (i = 1, \ldots, n).$$

Combining (5.19) with (5.15), we obtain the following interesting identity for the arithmetic function of $\langle a^{(0)} T^k \rangle$, $a^{(0)}$ from (5.17), since (5.15) can be written in the form

$$x_{i,k+n} = \sum_{r=1}^{n} (-1)^{r-1} \binom{n}{r} D^{n-r} x_{i,k+n-r} \qquad (i = 1, \ldots, n);$$

$$\sum_{j=0}^{n-i} \binom{n-i}{j} (D-1)^j f(n^2+nk+i-1+j)$$

(5.20)

$$= \sum_{r=1}^{n} (-1)^{r-1} \binom{n}{r} D^{n-r} \sum_{j=0}^{n-i} \binom{n-i}{j} (D-1)^j f(n^2+nk-nr+i-1+j)$$

$$(i = 1, \ldots, n).$$

Identities for the arithmetic functions of the GJPA can be obtained with the help of units that are of a more complex structure than that given by (5.4). We choose the unit given by (2.12) with the following restriction

(5.21) $\epsilon^{-1} = \frac{(w - D)^3}{k}$, $k|3D$, $w = \sqrt[3]{D^3 + k}$, $k > 1$.

We then obtain that, for $a^{(0)} = (w+2D, w^2+Dw+D^2)$, $\langle a^{(0)}T^k\rangle$ with the inner T-function is purely periodic with length of period 3. The unit obtained from the period equals, of course, $\epsilon = \frac{k}{(w - D)^3}$, and it is this unit which we now use for the above stated purpose.

Again let

$$\epsilon^{\ell} = \left(\frac{3}{(w - D)^3}\right)^{\ell}, \quad \epsilon^{\ell} = x_{\ell} + y_{\ell}w + z_{\ell}w^2,$$

$$(5.22) \quad \epsilon^{-1} = \frac{(w - D)^3}{k} = 1 + 3Dtw - 3tw^2,$$

$$D = kt, \quad w^3 = m = D^3 + k.$$

The reader will note that for $k|3D$, the coefficients of the powers of w in ϵ^{-1} are rational integers. Proceeding as before, we obtain

$$(1 + 3Dtw - 3tw^2)(x_{n+1} + y_{n+1}w + z_{n+1}w^2) = x_n + y_nw + z_n,$$

$$X_n = WX_{n+1}, \quad X_n = \begin{pmatrix} x_n \\ y_n \\ z_n \end{pmatrix},$$

$$W = \begin{pmatrix} 1 & -3tm & 3Dtm \\ 3Dt & 1 & -3tm \\ -3t & 3Dt & 1 \end{pmatrix},$$

and the wanted recurrence relation is given by

$$(5.23) \quad X_{n+3} - 3(1 + 9Dt^2m)X_{n+2} + 3X_{n+1} - X_n = Z.$$

Of course, one has to calculate the intital values for X_0, X_1, X_2. For X_1 we obtain

$$X_1' = (2D^4t^2 + Dt(2Dt+7tm)+1, \quad 3t^2(2m+D^3), \quad 3t(3D^3t+1)),$$

(X_1' stands for the transpose of X_1), while X_2 looks still much more complicated. Similar identities as in (5.20) are obtained from (5.23).

REFERENCES

1. L. Bernstein, The modified algorithm of Jacobi-Perron, Mem. Amer. Math. Soc. 67 (1966), 1-44.

2. L. Bernstein, Periodical continued fractions of degree n by Jacobi's algorithm, J. Reine. Angew. Math. 213 (1964), 31-38.

3. L. Bernstein, New infinite classes of periodic Jacobi-Perron algorithms, Pacific J. Math. 16 (1965), 1-31.

4. L. Bernstein, Einheitenberechnung in kubischen Koerpern mittels des Jacobi-Perronschen Algorithmus aus der Rechenanlage, J. Reine Angew. Math. 244 (1970), 201-220.

5. L. Bernstein, An explicit summation formula and its application, Proc. Amer. Math. Soc. 25, (1970), 323-334.

6. L. Bernstein, Units and fundamental units, (to appear),

7. L. Bernstein, The generalized Pellian equation, Trans. Amer. Math. Soc. 127 (1967), 76-89.

8. L. Bernstein and H. Hasse, Einheitenberechnung mittels des Jacobi-Perronschen Algorithmus, J. Reine Angew. Math. 218 (1965), 51-69.

9. L. Bernstein and H. Hasse, An explicit formula for the units of an algebraic number field of degree $n \geq 2$, Pacific J. Math. 30 (1969), 293-365.

10. L. Bernstein and H. Hasse, Explicit determination of the matrices in periodic algorithms of the Jacobi-Perron type with application to generalized Fibonacci numbers with time impulses, Fibonacci Quart. 7 (1969), 394-436.

11. M. Davis, An explicit Diophantine definition of the exponential function, (to appear).

12. C. G. J. Jacobi, Allgemeine Theory der kettenbruchaehnlichen Algorithmen, in welchen jede Zahl aus drei vorhergehenden gebildet wird, J. Reine Angew. Math. 69 (1869), 29-64.

13. Ju. V. Matijasevič, Enumerable sets are Diophantine (Russian), Soviet Math. Dokl. 11 (1970), 4-8.

14. O. Perron, Grundlagen fuer eine Theorie der Jacobischen Kettenbruchalgorithmus, Math. Ann. 64 (1907), 1-76.

15. J. Stender, Ueber die Grundeinheit fuer spezielle unendliche Klassen reiner kubischer Zahlkoerper, Abh. Math. Sem. Univ. Hamburg, 33 (1969), 203-215.

EULERIAN NUMBERS AND OPERATORS

L. Carlitz, Duke University

ABSTRACT

Put

(1) $$\frac{1-\lambda}{e^x-\lambda} = \sum_{n=0}^{\infty} H_n(\lambda)\frac{x^n}{n!}.$$

Then

$$H_n(\lambda) = (\lambda-1)^n \sum_{k=1}^{n} A_{n,k}\lambda^{k-1} \qquad (n \geq 1),$$

where the $A_{n,k}$ are positive integers that satisfy the recurrence

(2) $$A_{n+1,k} = (n-k+2)A_{n,k-1} + kA_{n,k}$$

and the symmetry condition

(3) $$A_{n,n-k+1} = A_{n,k}.$$

The first few values of $A_{n,k}$ are easily computed by means of the recurrence

n \ k	1	2	3	4	5
1	1				
2	1	1			
3	1	4	1		
4	1	11	11	1	
5	1	26	66	26	1

For references see [1].

Euler showed that

$$(4) \qquad A_{n,k} = \sum_{j=0}^{k} (-1)^j \binom{n+1}{j} (s-j)^n$$

while Worpitzky proved that

$$(5) \qquad x^n = \sum_{k=1}^{n} A_{n,k} \binom{x+k-1}{k}.$$

The latter formula may indeed be used to define the $A_{n,k}$.

To get a simple combinatorial interpretation of the $A_{n,k}$ we consider permutations of $Z_n = \{1, 2, \ldots, n\}$. Let $(a_1, a_2, \ldots, a_n)$ be a permutation of Z_n. A <u>rise</u> is a pair a_i, a_{i+1} with $a_i < a_{i+1}$; also it is customary to count a conventional rise to the left of a_1. For example the permutation (.2.51.43) has three rises as indicated by the dots. Let $P(n,k)$ denote the number of permutations of Z_n with k rises. By considering the effect of inserting the element n+1 in a permutation of Z_n with k rises we get the recurrence

$$P(n+1,k) = (n-k+2)P(n,k-1) + kP(n,k).$$

Since this recurrence is identical with (2) and the intial values of $A_{n,k}$ and $P(n,k)$ are the same, it follows that

$$(6) \qquad P(n,k) = A_{n,k}.$$

Since the symmetry condition (3) is by no means apparent from the generating function (1), it seems desirable to introduce some new notation. Put

$$(7) \qquad A(r,s) = A_{r+s+1,r+1} = A_{r+s+1,s+1} = A(s,r).$$

It is found that (1) implies

$$(8) \qquad F(x,y) \equiv \frac{e^x - e^y}{xe^y - ye^y} = \sum_{r,s=0}^{\infty} A(r,s) \frac{x^r y^s}{(r+s+1)!}.$$

Moreover since

$$(D_x + D_y)F = F^2,$$

where

$$D_x = \partial/\partial x, \quad D_y = \partial/\partial y,$$

it follows from (8) that

$$\sum_{r,s=0}^{\infty} A(r,s)\frac{x^r y^s}{(r+s)!} = (1 + xF)(1 + yF). \tag{9}$$

The generating function (9) suggests the generalization

$$\sum_{r,s=0}^{\infty} A(r,s|\alpha,\beta)\frac{x^r y^s}{(r+s)!} = (1 + xF)^{\alpha}(1 + yF)^{\beta}. \tag{10}$$

Clearly

$$A(r,s|1,1) = A(r,s).$$

It follows from (10) that

$$A(r,s|\alpha,\beta) = (r+\beta)A(r,s-1|\alpha,\beta) + (s+\alpha)A(r-1,s|\alpha,\beta) \tag{11}$$

and

$$A(r,s|\alpha,\beta) = A(s,r|\beta,\alpha). \tag{12}$$

Moreover

$$A(r,s|\alpha,\beta) = \sum_{j=0}^{r} (-1)^{r-j}\binom{\alpha+\beta+j-1}{j}\binom{\alpha+\beta+r+s}{r-j}(\beta+j)^{r+s} \tag{13}$$

and

$$(\alpha+x)^r\binom{\alpha+\beta+x-1}{\alpha+\beta-1} = \sum_{s=0}^{r}\binom{\alpha+\beta+x+s-1}{\alpha+\beta+r-1}A(s,r-s|\alpha,\beta), \tag{14}$$

when $\alpha+\beta$ is a positive integer. These formulas generalize (4) and (5), respectively.

The $A(r,s|\alpha,\beta)$ admit of the following combinatorial interpretation when α,β are positive integers. Let $\alpha = (a_1, a_2, \ldots, a_n)$ be a permutation of Z_n with k rises. Let $0 \le j \le m$ and let $C(n, k; m, j)$ denote the number of ways of inserting $n+1, \ldots, n+m$ in α so that the resulting permutation of Z_{n+m} will have exactly $k+j$ rises. Then

$$C(n, k; m, j) = A(j,m-j|n-k+1,k). \tag{15}$$

Put

$$A_n(x,y) = \sum_{r+s=n} A(r,s)x^r y^s.$$

It follows from (2) that

$$A_{n+1}(x,y) = [x+y+xy(D_x + D_y)]\, A_n(x,y), \tag{16}$$

so that

$$A_n(x,y) = [x+y+xy(D_x + D_y)]^n \cdot 1. \tag{17}$$

Moreover the operator

$$\Omega^n = [x+y+xy(D_x + D_y)]^n$$

can be expanded in the following way.

$$\Omega^n = \sum_{k=0}^{n} \frac{(xy)^k}{k!\,(k+1)!} (D_x + D_y)^k A_n(x,y) \cdot (D_x + D_y)^k \tag{18}$$

It follows from (17) and (18) that

$$A_{m+n}(x,y) = \sum_{k=0}^{\min(m,n)} \frac{(xy)^k}{k!\,(k+1)!} (D_x + D_y)^k A_m(x,y) \cdot (D_x + D_y)^k A_n(x,y). \tag{19}$$

The generating function (10) suggests that we define

$$A_n(x,y|\alpha,\beta) = \sum_{r+s=n} A(r,s|\alpha,\beta)x^r y^s. \tag{20}$$

It follows from (11) that

$$A_{n+1}(x,y|\alpha,\beta) = \Omega_{\alpha,\beta} A_n(x,y|\alpha,\beta), \tag{21}$$

where

$$\Omega_{\alpha,\beta} \equiv \alpha x + \beta y + xy(D_x + D_y). \tag{22}$$

Thus

$$A_n(x,y|\alpha,\beta) = \Omega^n_{\alpha,\beta}\cdot 1. \tag{23}$$

Corresponding to (18) we have

$$\Omega^n_{\alpha,\beta} = \sum_{k=0}^{n} \frac{(xy)^k}{k!(\alpha+\beta)_k}(D_x + D_y)^k A_n(x,y|\alpha,\beta)\cdot(D_x + D_y)^k, \tag{24}$$

where

$$(\alpha+\beta)_k = (\alpha+\beta)(\alpha+\beta+1)\cdots(\alpha+\beta+k-1).$$

It follows from (23) and (24) that

$$A_{m+n}(x,y|\alpha,\beta)$$

$$= \sum_{k=0}^{\min(m,n)} \frac{(xy)^k}{k!(\alpha+\beta)_k}(D_x + D_y)^k A_m(x,y|\alpha,\beta)\cdot(D_x + D_y)^k A_n(x,y|\alpha,\beta). \tag{25}$$

REFERENCES

1. L. Carlitz, Eulerian numbers and polynomials, Math. Mag. 33 (1959), 247-260.

2. L. Carlitz, Eulerian numbers and operators, in preparation.

3. L. Euler, Institutiones Calculi Differentialis, Omnia Opera (1), Vol. 10 (1913).

4. D. Foata and M. P. Schutzenberger, Theorie Geometrique des Polynomes Euleriens, Springer, Berlin-Heidelberg-New York, 1970.

5. G. Frobenius, Über die Bernoullischen Zahlen und die Eulerschen Polynome, S.-B. Preuss. Akad. Wiss. (1910), 809-847.

6. J. Riordan, An Introduction to Combinatorial Analysis, Wiley, New York, 1958.

7. J. Worpitzky, Über die Bernoullischen und Eulerschen Zahlen, J. Reine Angew. Math. 94 (1883), 203-232.

SOME FUNCTIONS RELATED TO FIBONACCI AND LUCAS REPRESENTATIONS

L. Carlitz*, Duke University
V. E. Hoggatt, San Jose State College
Richard Scoville, Duke University

1. Introduction. In the usual notation for Fibonacci numbers, put

$$F_0 = 0, \quad F_1 = 1, \quad F_{n+1} = F_n + F_{n-1} \quad (n \geq 1).$$

It is well known that any positive integer N is uniquely representable in the form

(1.1) $$N = F_{k_1} + F_{k_2} + \cdots + F_{k_t},$$

where

(1.2) $$k_1 \geq 2, \quad k_{i+1} - k_i \geq 2 \quad (i = 1, 2, \ldots, t-1).$$

We shall refer to (1.1) as the canonical representation. Note that $t = t(N)$ is a function of N.

If N satisfies (1.1), where

(1.3) $$1 < k_1 < k_2 < \cdots < k_t,$$

then we call (1.1) a Fibonacci representation. The number of such representations is denoted by $R(N)$; if, however, t is held fixed, the number of representations is denoted by $R_t(N)$. The functions $R(N)$ and $R_t(N)$ have been discussed in [2] and [3] as well as in several other papers referred to in [2] and [3]. In particular, recurrence formulas for these functions were obtained. Considerable use was made of the function $e(N)$ defined by (1.1) together with

(1.4) $$e(N) = F_{k_1-1} + F_{k_2-1} + \cdots + F_{k_t-1};$$

it should be noted that the definition (1.4) is independent of the particular representation (1.1).

Let A_k denote the set of positive integers N with $k_1 = k$ in the canonical representation (1.1). Thus the

*Supported in part by NSF grant GP-17031.

(1.5) $A_k \quad (k = 2, 3, 4, \ldots)$

constitute a partition of the positive integers. To identify the numbers in A_k we make use of the functions

(1.6) $a(n) = [\alpha n], \quad b(n) = [\alpha^2 n] = a(n) + n,$

where

$$\alpha = \tfrac{1}{2}(1 + \sqrt{5})$$

and $[x]$ denotes the greatest integer function. Then we have the following results:

(1.7) $A_{2t} = \{ab^{t-1}a(n) \mid n = 1, 2, 3, \ldots\} \quad (t = 1, 2, 3, \ldots)$

and

(1.8) $A_{2t+1} = \{b^t a(n) \mid n = 1, 2, 3, \ldots\} \quad (t = 1, 2, 3, \ldots).$

It follows at once from the definition that if $N \in A_t$, $t > 2$, then $e(n) \in A_{t-1}$. Also it is proved in [1] that

(i) for every N, $e(N+1) \geq e(N)$ with equality if and only if $N \in A_2$,

(ii) if $N \in A_2$ then neither $N-1$ nor $N+1$ is in A_2.

The first part of the present paper is concerned with Fibonacci representations. In the second part we consider similar problems for the Lucas numbers. The third part is concerned with Fibonacci numbers of the third order defined by

(1.9) $G_0 = 0, \quad G_1 = G_2 = 1, \quad G_{n+1} = G_n + G_{n-1} + G_{n-2} \quad (n \geq 2).$

In the fourth part we discuss briefly Pellian representations. The numbers in question are defined by

(1.10) $P_0 = 0, \quad P_1 = 1, \quad P_{n+1} = 2P_n + P_{n-1} \quad (n \geq 1).$

Finally, in the fifth part, we consider the sequence defined by

(1.11) $u_0 = 0, \quad u_1 = 1, \quad u_{n+1} = u_n + 2u_{n-1} \quad (n \geq 1).$

Unlike the other sequences, the explicit formula for u_n:

$$u_n = \frac{1}{3}\left(2^n - (-1)^n\right)$$

does not involve any irrationalities. Despite this, many of the results obtained for the u_n are similar to those obtained for the other sequences.

The present paper lists theorems but contains no proofs. A detailed treatment is contained in the papers [4], ..., [9].

I. FIBONACCI REPRESENTATIONS

2. The array R. As in [1] we form the 3-rowed array R in the following way. In the first row we place the positive integers in natural order. We begin the second row with the entry 1. To get an entry of the third row we add the entries appearing above in the first and second rows. We get further entries in the second row by choosing the smallest integer which has not appeared so far in the second or third rows.

R:

1	2	3	4	5	6	7	8	9	10	...
1	3	4	6	8	9	11	12	14	16	...
2	5	7	10	13	15	18	20	23	26	...

The array R is uniquely determined by the following properties.

(i) every positive integer appears exactly once in row 2 or row 3,

(ii) each row is a monotone sequence,

(iii) the sum of the first two rows is the third row.

Next consider the 3-rowed array

R':

1	2	3	4	5	...
a(1)	a(2)	a(3)	a(4)	a(5)	...
b(1)	b(2)	b(3)	b(4)	b(5)	...

where $a(n)$, $b(n)$ are defined by (1.6). Then we have

<u>Theorem</u> 2.1. <u>The arrays</u> R <u>and</u> R' <u>are identical</u>.

Making use of the properties (i), (ii) and (iii) of the array R, we obtain a number of relations involving the functions $a(n)$, $b(n)$:

(2.1) $n + a(n) = b(n)$

(2.2) $b(n) = a^2(n) + 1$

(2.3) $a(n) + b(n) = ba(n) + 1$

(2.4) $ab(n) = ba(n) + 1$

(2.5) $a(n) + b(n) = ab(n)$

(2.6) $b^2(n) = aba(n) + 2$

(2.7) $ab^2(n) = b^2a(n) + 3$

(2.8) $b^r(n) = ab^{r-1}a(n) + F_{2r-1} \quad (r = 1, 2, 3, \ldots)$

(2.9) $ab^r(n) = b^ra(n) + F_{2r} \quad (r = 1, 2, 3, \ldots)$

(2.10) $b^r(1) = F_{2r+1} \quad (r = 1, 2, 3, \ldots).$

Here of course

$$ab(n) = a(b(n)), \quad a^2(n) = a(a(n)),$$

and so on.

3. <u>The sets</u> A_k.

<u>Theorem</u> 3.1. <u>If</u> $N \in A_2$, <u>then</u> $N+1 \in A_k$ <u>with</u> k <u>odd</u>.

<u>Theorem</u> 3.2. <u>If</u> M <u>and</u> N <u>are in</u> A_2 <u>and</u> $e^2(M) = e^2(N)$, <u>then</u> $M = N$.

<u>Theorem</u> 3.3. <u>Put</u>

(3.1) $$A_2 = \{Q_1, Q_2, Q_3, \ldots\},$$

<u>where</u>

$$Q_1 < Q_2 < Q_3 < \cdots.$$

<u>Then</u>

$$e^2(Q_j) = j.$$

Theorem 3.4. *We have*

$$e(a(N)) = N, \quad e(b(N)) = a(N) \qquad (N = 1, 2, 3, \ldots).$$

Define N_j by means of

$$e(N_j) = j, \quad e(N_j - 1) \neq j, \tag{3.2}$$

where $e(0) = 0$. Thus $N_1 = 1$, $N_2 = 3$, etc.

Theorem 3.5. *The numbers* $(N_1, N_2, \ldots)$ *and* $(Q_1+1, Q_2+1, \ldots)$ *constitute the second and third rows of the array* R.

Theorem 3.6. *We have*

$$A_{2t} = \{ab^{t-1}a(n) \mid n = 1, 2, 3, \ldots\} \qquad (t = 1, 2, 3, \ldots),$$

$$A_{2t+1} = \{b^t a(n) \mid n = 1, 2, 3, \ldots\} \qquad (t = 1, 2, 3, \ldots).$$

Let $A(s, s+2)$ denote the set of positive integers with canonical representation

$$F_{k_1} + F_{k_2} + \cdots + F_{k_t}$$

satisfying

$$k_1 = s, \quad k_2 = s + 2$$

and let $A(s, \overline{s+2})$ denote the set of positive integers for which

$$k_1 = s, \quad k_2 > s + 2.$$

Theorem 3.7. *For* $t \geq 1$ *we have*

$$A(2t, 2t+2) = \{ab^{t-1}a^2(n) \mid n = 1, 2, 3, \ldots\},$$

$$A(2t, 2t+2) = \{ab^{t-1}ab(n) \mid n = 1, 2, 3, \ldots\},$$

$$A(2t+1, \overline{2t+3}) = \{b^t a^2(n) \mid n = 1, 2, 3, \ldots\},$$

$$A(2t+1, 2t+3) = \{b^t ab(n) \mid n = 1, 2, 3, \ldots\}.$$

4. Some additional functions. We define the function $\lambda(n)$ by means of $\lambda(1) = 0$ and $\lambda(n) = t$, where $n > 1$ and t is the smallest integer such that

$$e^t(n) = 1. \tag{4.1}$$

Theorem 4.1. Let

$$N = F_{k_1} + F_{k_2} + \cdots + F_{k_r}$$

be the canonical representation of N, so that

$$k_1 \geq 2, \quad k_{i+1} - k_i \geq 2 \quad (i = 1, 2, \ldots, r-1).$$

Then

$$\lambda(N) = \begin{cases} k_r - 2 & (r = 1) \\ k_r - 1 & (r > 1). \end{cases} \tag{4.2}$$

Next let Λ_t denote the set of positive integers N such that

$$\lambda(N) = t. \tag{4.3}$$

Theorem 4.2. Λ_t consists of the integers N such that

$$F_{t+1} < N \leq F_{t+2}. \tag{4.4}$$

Hence

$$\text{card } \Lambda_t = F_t. \tag{4.5}$$

Theorem 4.3. Let

$$\{x\} = x - [x]$$

denote the fractional part of the real number x. Then

$$n \in a(\underset{\sim}{N}) \Leftrightarrow 0 < \left\{\frac{N}{\alpha^2}\right\} < \frac{1}{\alpha}, \tag{4.6}$$

$$n \in b(\underset{\sim}{N}) \Leftrightarrow \frac{1}{\alpha} < \left\{\frac{N}{\alpha^2}\right\} < 1, \tag{4.7}$$

where $\underset{\sim}{N}$ denotes the set of positive integers.

5. _Word functions_. By a _word function_ (or briefly a _word_) is meant any monomial in the a's and b's. It is convenient to include 1 as a word. If u, v are any words, then $au \neq bv$. Also if $au = av$ or $bu = bv$ then $u = v$. It follows that any word is uniquely representable as a product of "primes" a, b.

We define the _weight_ of a word by means of

$$p(1) = 0, \quad p(a) = 1, \quad p(b) = 2 \tag{5.1}$$

together with

$$p(uv) = p(u) + p(v). \tag{5.2}$$

It follows that there is exactly one word of weight 1, two of weight 2 and three of weight 3. Let N_p denote the number of words of weight p. It follows readily that

$$N_p = N_{p-1} + N_{p-2} \quad (p \geq 2).$$

Since $N_0 = N_1 = 1$, it is evident that

$$N_p = F_{p+1} \quad (p \geq 0). \tag{5.3}$$

We next inquire when the words u, v satisfy

$$uv = vu. \tag{5.4}$$

Theorem 5.1. _The words_ u, v _satisfy (5.4) if and only if there is a word_ w _such that_

$$u = w^r, \quad v = w^s,$$

where r, s _are nonnegative integers_.

We show next that any word is "almost" linear. More precisely we have

Theorem 5.2. _Any word_ u _of weight_ p _is uniquely representable in the form_

$$u(n) = F_p \cdot a(n) + F_{p-1} \cdot n - \lambda_u, \tag{5.5}$$

where λ_u _is independent of_ n. _Moreover_

(5.6) $$\lambda_{ua} = \lambda_u + F_p, \quad \lambda_{ub} = \lambda_u,$$

where u *is of weight* p.

As a corollary of Theorem 5.2 we have

Theorem 5.3. *For arbitrary* u, v, *we have*

(5.7) $$uv - vu = C,$$

where C *is independent of* n.

A few special cases of (5.5) follow:

(5.8) $$a^k(n) = F_k a(n) + F_{k-1} n - F_{k+1} + 1,$$

(5.9) $$b^k(n) = F_{2k} a(n) + F_{2k-1} n,$$

(5.10) $$b^k(n) = a^{2k}(n) + F_{2k+1} - 1,$$

(5.11) $$(ab)^k(n) = F_{3k} a(n) + F_{3k-1} - \tfrac{1}{2}(F_{3k-1} - 1),$$

(5.12) $$(ba)^k(n) = F_{3k} a(n) + F_{3k-1}(n) - F_{3k-1},$$

(5.13) $$(ab)^k(n) - (ba)^k(n) = \tfrac{1}{2}(F_{3k-1} + 1).$$

6. *Generating functions*. Put

(6.1) $$\varphi_j(x) = \sum_{n \in A_j} x^n \quad (j = 2, 3, 4, \ldots).$$

It follows at once that

(6.2) $$\frac{x}{1 - x} = \sum_{j=2}^{\infty} \varphi_j(x).$$

Also from the definition of A_r it is clear that

(6.3) $$\varphi_r(x) = x^{F_r}\left\{1 + \sum_{j=r+2}^{\infty} \varphi_j(x)\right\} \quad (r = 2, 3, 4, \ldots).$$

This implies

(6.4) $$x^{-F_r}\varphi_r(x) - x^{F_{r+1}}\varphi_{r+1}(x) = \varphi_{r+2}(x) \qquad (r = 2, 3, 4, \ldots).$$

For $r = 2$, (6.3) reduces to

$$\varphi_2(x) = x\left\{1 + \sum_{j=4}^{\infty} \varphi_j(x)\right\};$$

combining this with (6.2) we get

(6.5) $$(1 + x)\varphi_2(x) + x\varphi_3(x) = \frac{x}{1 - x}.$$

It is convenient to define

(6.6) $$\varphi(x) = \sum_{n=1}^{\infty} x^{a(n)}.$$

It follows that

(6.7) $$\varphi(x) = \varphi_2(x) + x\varphi_3(x).$$

Moreover

(6.8) $$x\,\varphi_2(x) = \frac{x}{1 - x} - \varphi(x), \quad x^2\varphi_3(x) = -\frac{x}{1 - x} + (1 + x)\varphi(x).$$

Generally we have

(6.9) $$x^{F_{r+1}-1}\varphi_r(x) = (-1)^r\left\{\frac{xA_r(x)}{1 - x} - B_r(x)\varphi(x)\right\},$$

where

$$xA_r(x) = \sum_{j=1}^{F_{r-1}} x^{a(j)}, \quad B_r(x) = \frac{1 - x^{F_r}}{1 - x}.$$

<u>Theorem</u> 6.1. <u>The function</u> $\varphi(x)$ <u>possesses the unit circle as a natural boundary</u>.

In view of (6.9), the theorem holds for every $\varphi_r(x)$.

7. Second canonical representation.

Theorem 7.1. Every positive integer N is uniquely representable in the form

$$N = F_{k_1} + F_{k_2} + \cdots + F_{k_r}, \tag{7.1}$$

where k_1 is odd and

$$k_{i+1} - k_i \geq 2 \quad (i = 1, 2, \ldots, r-1).$$

Note that in (7.1) the value $k_1 = 1$ is allowed.

In view of Theorem 7.1 we let $\bar{A}_{2k+1}$ denote the set of integers $\{N\}$ with $k_1 = 2k+1$. Then the sets

$$\bar{A}_{2k+1} \quad (k = 0, 1, 2, \ldots)$$

constitute a partition of the positive integers. Clearly

$$\bar{A}_{2k+1} = A_{2k+1} \quad (k = 1, 2, 3, \ldots) \tag{7.2}$$

and

$$\bar{A}_1 = \bigcup_{t=1}^{\infty} A_{2t} = a(\underset{\sim}{N}), \tag{7.3}$$

where as above $\underset{\sim}{N}$ denotes the set of positive integers.

In (7.1) take $k_1 = 1$, so that $N \in \bar{A}_1$. If $k_2 > 3$ we find that $N \in A_2$. If N has the second canonical representation

$$N = F_1 + F_3 + \cdots + F_{2s-1} + F_{k_1} + F_{k_2} + \cdots,$$

where

$$k_{j+1} - k_j \geq 2 \quad (j \geq 1), \quad k_1 \geq 2s + 2,$$

then $N \in A_{2s}$ and conversely.

II. LUCAS REPRESENTATIONS

8. Canonical representation. We define the Lucas numbers as usual by

$$L_0 = 2, \quad L_1 = 1, \quad L_{n+1} = L_n + L_{n-1} \quad (n \geq 1).$$

Theorem 8.1. Every positive integer N is uniquely representable either in the form

(8.1) $$N = L_0 + L_{k_1} + \cdots + L_{k_r},$$

where

(8.2) $$k_{i+1} - k_i \geq 2 \quad (i = 1, 2, \ldots, r-1), \quad k_1 \geq 3,$$

or in the form

(8.3) $$N = L_{k_1} + \cdots + L_{k_r},$$

where

(8.4) $$k_{i+1} - k_i \geq 2 \quad (i = 1, 2, \ldots, r-1), \quad k_1 \geq 1.$$

No integer is representable in both (8.1) and (8.3).

Let P_n denote the set of integers that can be written in the form (8.1) with $k_r \leq n$ and let Q_n be the set of integers that can be written in the form (8.3) with $k_r \leq n$. Thus

$$P_3 = \{2, 6\}, \quad Q_3 = \{1, 3, 4, 5\},$$

$$P_4 = \{2, 6, 9\}, \quad Q_4 = \{1, 3, 4, 5, 7, 8, 10\}.$$

Theorem 8.2. We have

(8.5) $$P_n \cup Q_n = \{1, 2, 3, \ldots, L_{n+1} - 1\},$$

(8.6) $$\text{card } (P_n) = F_n,$$

(8.7) $$\text{card } (Q_n) = F_{n+2} - 1.$$

Let B_0 denote the set of positive integers representable in the form (8.1) and let B_k denote the set of positive integers representable in the form (8.3) with $k_1 = k$, $k \geq 1$. Then the sets

$$B_k \quad (k = 0, 1, 2, \ldots) \tag{8.8}$$

constitute a partition of the positive integers. This may be called a *Lucas partition*.

Corresponding to Theorem 3.6, we have

Theorem 8.3. *The following relations hold:*

$$B_0 = \{a^2(n) + n \mid n = 1, 2, 3, \ldots\}, \tag{8.9}$$

$$B_1 = \{a^2(n) + n - 1 \mid n = 1, 2, 3, \ldots\}, \tag{8.10}$$

$$B_{2t} = \{b^{t-1}a(n) + b^t a(n) \mid n = 1, 2, 3, \ldots\} \quad (t = 1, 2, 3, \ldots), \tag{8.11}$$

$$B_{2t+1} = \{ab^{t-1}a(n) + ab^t a(n) \mid n = 1, 2, 3, \ldots\} \quad (t = 1, 2, 3, \ldots). \tag{8.12}$$

9. *Application to proof of conjectures.*

Theorem 9.1. *An integer* N *is in* B_0 *if and only if it is not representable in the form*

$$N = L_{j_1} + L_{j_2} + \cdots + L_{j_s} \tag{9.1}$$

with

$$1 \leq j_1 < j_2 < \cdots < j_s.$$

Now let $\nu(n)$ denote the number of positive integers $N \leq n$ that are not representable in the form (9.1), so that by Theorem 9.1, $\nu(n)$ is also the number of integers $\leq n$ in B_0.

Theorem 9.2. *We have*

$$\nu(n) = \left[\frac{n+2}{\alpha^2 + 1}\right] \quad (\alpha = \tfrac{1}{2}(1 + \sqrt{5})), \tag{9.2}$$

where $[x]$ *is the greatest integer function.*

Theorem 9.3. We have

$$\nu(L_n) = F_{n-1} \quad (n \geq 1). \tag{9.3}$$

This result was conjectured by Hoggatt and proved by Klarner. More generally we have the following results which were also conjectured by Hoggatt.

Theorem 9.4. Let k be a fixed positive integer. Then

$$\nu(kL_n) = kF_{n-1} \tag{9.4}$$

for n sufficiently large.

Theorem 9.5. We have

$$\nu(5F_n) = L_{n-1} \quad (n > 1) \tag{9.5}$$

and

$$\nu(5kF_n) = k\, L_{n-1} \tag{9.6}$$

for n sufficiently large.

10. Generating functions. Put

$$\psi_j(x) = \sum_{n \in B_j} x^n \quad (j = 0, 1, 2, \ldots). \tag{10.1}$$

Clearly, by (8.9) and (8.10),

$$\psi_0(x) = x\, \psi_1(x); \tag{10.2}$$

also

$$\frac{x}{1-x} = \sum_{j=0}^{\infty} \psi_j(x), \tag{10.3}$$

$$\frac{x}{1-x} = (1+x)\psi_1(x) + \sum_{j=2}^{\infty} \psi_j(x). \tag{10.4}$$

In the next place it follows from the definition of B_r that

$$\psi_r(x) = x^{L_r}\left\{1 + \sum_{j=r+2}^{\infty} \psi_j(x)\right\} \quad (r \geq 1). \tag{10.5}$$

This implies

$$x^{-L_r}\psi_r(x) - x^{-L_{r+1}}\psi_{r+1}(x) = \psi_{r+2}(x) \quad (r \geq 1). \tag{10.6}$$

In particular

$$\psi_1(x) = x\left\{1 + \sum_{j=3}^{\infty} \psi_j(x)\right\}, \tag{10.7}$$

so that, by (10.4),

$$\frac{x}{1-x} = (1 + x + x^2)\psi_1(x) + x\psi_2(x). \tag{10.8}$$

By means of (10.6) and (10.8) we can express all $\psi_j(x)$, $j > 1$, in terms of $\psi_1(x)$. For example

$$x\psi_2(x) = \frac{x}{1-x} - (1 + x + x^2)\psi_1(x),$$

$$x^4\psi_3(x) = -\frac{x}{1-x} + (1 + x + x^2 + x^3)\psi_1(x),$$

$$x^8\psi_4(x) = \frac{x + x^5}{1-x} - \frac{1-x^7}{1-x}\psi_1(x).$$

Generally we have

$$x^{L_{r+1}-3}\psi_r(x) = (-1)^r\left\{\frac{xA_r(x)}{1-x} - B_r(x)\psi_1(x)\right\}, \tag{10.9}$$

where

$$xA_r(x) = \sum_{n=1}^{F'_{r-1}} x^{a^2(n)+n-1}, \quad B_r(x) = \frac{1-x^{L_r}}{1-x}. \tag{10.10}$$

Theorem 10.1. The unit circle is the natural boundary of the function $\psi_1(x)$.

In view of (10.9) this theorem holds for each of the functions $\psi_k(x)$.

It would be of interest to know whether there is any simple relation connecting $\psi_1(x)$ with

$$\varphi(x) = \sum_{n=1}^{\infty} x^{a(n)},$$

the function introduced in Section 6. In particular do there exist polynomials $P(x)$, $Q(x)$, $R(x)$ such that

$$P(x)\varphi(x) + Q(x)\psi_1(x) = R(x)? \tag{10.11}$$

III. FIBONACCI REPRESENTATIONS OF HIGHER ORDER

11. Canonical representation. We define the Fibonacci numbers of the third order by means of

$$G_0 = 0, \quad G_1 = G_2 = 1, \quad G_{n+1} = G_n + G_{n-1} + G_{n-2} \quad (n \ge 2). \tag{11.1}$$

Theorem 11.1. Any positive integer can be uniquely represented in the form

$$N = \epsilon_2 G_2 + \epsilon_3 G_3 + \epsilon_4 G_4 + \cdots, \tag{11.2}$$

where each $\epsilon_i = 0$ or 1 and

$$\epsilon_i \epsilon_{i+1} \epsilon_{i+2} = 0 \quad (i = 2, 3, 4, \ldots). \tag{11.3}$$

When (11.3) is satisfied we call (11.2) the canonical representation of N.

If N has the canonical representation (11.2) we introduce the function

$$f(N) = \epsilon_2 G_1 + \epsilon_3 G_2 + \epsilon_4 G_3 + \cdots . \tag{11.4}$$

Theorem 11.2. Let

$$N = \epsilon_2' G_2 + \epsilon_3' G_3 + \epsilon_4' G_4 + \cdots,$$

where each $\epsilon_i = 0$ or 1, be any representation of N. Then

$$f(N) = \epsilon_2' G_1 + \epsilon_3' G_2 + \epsilon_4' G_3 + \cdots . \tag{11.5}$$

Let C_k denote the set of positive integers $\{N\}$ for which ϵ_k is the first nonvanishing ϵ_i in the canonical representation (11.2).

Theorem 11.3. We have $f(N+1) \geq f(N)$, with equality if and only if $N \in C_2$.

Theorem 11.4. We have $N-1 \notin C_2$ if and only if $N \in C_k$, where $k \equiv 2 \pmod 3$.

Theorem 11.5. The following identities hold for $k > 2$.

$$f(G_k - 1) = \begin{cases} G_{k-1} & (k \equiv 0 \pmod 3), \\ G_{k-1} & (k \equiv 1 \pmod 3), \\ G_{k-1} - 1 & (k \equiv 2 \pmod 3). \end{cases} \tag{11.6}$$

12. The functions a, b, c. We define three strictly monotone functions on the positive integers which we display as an array.

R:

1	2	3	4	5	...
a(1)	a(2)	a(3)	a(4)	a(5)	...
b(1)	b(2)	b(3)	b(4)	b(5)	...
c(1)	c(2)	c(3)	c(4)	c(5)	...

We set $a(1) = 1$, $b(1) = 2$, $c(1) = 4$, $a(2) = 3$ and fill the rest of the array recursively. Suppose that columns 1 to n have been filled and also that $a(n+1)$ is known. Then we fill row a to column $a(n+1)$ in increasing order with the first integers that have not appeared so far in the array. Then we let $b(n+1)$ be the next integer that has not appeared and put

$$c(n+1) = n + 1 + a(n+1) + b(n+1).$$

Thus we get the following entries

R:

n	1	2	3	4	5	6	7	8	9	10
a	1	3	5	7	8	10	12	14	16	18
b	2	6	9	13	15	19	22	26	30	
c	4	11	17	24	28	35	41	48	55	

It follows from the definition of R that the ranges a(N), b(N), c(N) are disjoint and exhaust the positive integers.

Theorem 12.1. The following identities hold:

(12.1) $$c(N) = a(N) + b(N) + N,$$

(12.2) $$b(N) = a^2(N) + 1,$$

(12.3) $$ab(N) = ba(N) + 1,$$

(12.4) $$c(N) = ab(N) + 1 = ba(N) + 2.$$

Theorem 12.2. Let a_1, b_1, c_1 be strictly monotone functions whose ranges form a disjoint partition of the positive integers. Assume also that they satisfy

(12.5) $$c_1(N) = a_1(N) + b_1(N) + n,$$

(12.6) $$b_1(N) = a_1^2(N) + 1.$$

Then $a_1 = a$, $b_1 = b$, $c_1 = c$.

13. Some properties of f. Since f is monotone and every number appears in the range of f, the following definition is meaningful. For every N, let A(N) be defined by

(13.1) $$f(A(N)) = N, \quad f(A(N)-1) = N - 1.$$

Next define B(N) by

(13.2) $$B(N) = A(A(N)) + 1$$

and C(N) by

(13.3) $$C(N) = A(N) + B(N) + N.$$

Theorem 13.1. We have

$$A(A(\underset{\sim}{N})) \subseteq C_2, \tag{13.4}$$

where $\underset{\sim}{N}$ denotes the set of positive integers.

Theorem 13.2. We have

$$C_2 = A(A(\underset{\sim}{N})) \cup B(\underset{\sim}{N}). \tag{13.5}$$

Theorem 13.3. Let $K > 1$ be arbitrary and let $K-1$ be given canonically by

$$K - 1 = \epsilon_2 G_2 + \epsilon_3 G_3 + \epsilon_4 G_4 + \cdots . \tag{13.6}$$

Then

$$A(K) = G_1 + \epsilon_2 G_3 + \epsilon_3 G_4 + \cdots, \tag{13.7}$$

$$A(A(K)) = G_2 + \epsilon_2 G_4 + \epsilon_3 G_5 + \cdots, \tag{13.8}$$

$$C(K) = G_4 + \epsilon_2 G_5 + \epsilon_3 G_6 + \cdots . \tag{13.9}$$

Theorem 13.4. We have

$$A = a, \quad B = b, \quad C = c. \tag{13.10}$$

Moreover

$$\begin{cases} f(a(N)) = N, \\ f(b(N)) = a(N), \\ f(c(N)) = b(N). \end{cases} \tag{13.11}$$

14. The second canonical representation.

Theorem 14.1. Every positive integer N can be uniquely represented in the form

$$N = G_{3s+1} + \epsilon_{3s+2} G_{3s+2} + \epsilon_{3s+3} G_{3s+3} + \cdots, \tag{14.1}$$

where $s \ge 0$ and

$$\epsilon_i \epsilon_{i+1} \epsilon_{i+2} = 0 \quad (\epsilon_{3s+1} = 1). \tag{14.2}$$

Moreover

(14.3) $$a(N) = G_{3s+2} + \epsilon_{3s+2}G_{3s+3} + \cdots,$$

(14.4) $$b(N) = G_{3s+3} + \epsilon_{3s+2}G_{3s+4} + \cdots,$$

(14.5) $$c(N) = G_{3s+4} + \epsilon_{3s+2}G_{3s+5} + \cdots .$$

We may call (14.1) the second canonical representation.

In view of Theorem 14.1 we let $\overline{C}_{3s+1}$ denote the set of integers $\{N\}$ representable in the form (14.1). Clearly

(14.6) $$\overline{C}_{3s+1} = C_{3s+1} \quad (s \ge 1)$$

and

(14.7) $$\overline{C}_1 = \bigcup_{k=0}^{\infty} (C_{3k+2} \cup C_{3k+3}).$$

Theorem 14.2. The following formulas hold.

(14.8) $$\begin{aligned} a^2 &= b - 1, \\ ab &= c - 1, \\ ac &= a + b + c, \\ ba &= c - 2, \\ b^2 &= a + b + c - 1, \\ bc &= a + 2b + 2c , \\ ca &= a + b + c - 3, \\ cb &= a + 2b + 2c - 2, \\ c^2 &= 2a + 3b + 4c. \end{aligned}$$

A word function (or briefly word) is a monomial in a, b, c:

(14.9) $$u = a^{i_1} b^{j_1} c^{k_1} \cdots a^{i_r} b^{j_r} c^{k_r},$$

where the exponents are arbitrary nonnegative integers. Since (i) $au = bv$, $au = cw$, $bv = cw$, where u, v, w are arbitrary words, are all impossible and (ii) any one of $au = av$, $bu = bv$, $cu = cv$ implies $u = v$, it follows that the representation (14.9) is unique.

We define the <u>weight</u> of a word by means of

$$p(1) = 0, \quad p(a) = 1, \quad p(b) = 2, \quad p(c) = 3$$

together with

$$p(uv) = p(u) + p(v).$$

Let N_p denote the number of words of weight p. Then we have

(14.10) $$N_p = N_{p-1} + N_{p-2} + N_{p-3} \quad (p \geq 3).$$

Since

$$N_0 = N_1 = 1, \quad N_2 = 2,$$

it follows that

(14.11) $$N_p = G_{p+1} \quad (p \geq 0).$$

<u>Theorem</u> 14.3. <u>The words</u> u, v <u>satisfy</u>

(14.12) $$uv = vu$$

<u>if and only if there is a word</u> w <u>such that</u>

$$u = w^r, \quad v = w^s$$

<u>where</u> r, s <u>are nonnegative integers</u>.

<u>Theorem</u> 14.4. <u>Let</u> u <u>be a word of weight</u> p. <u>Then</u>

(14.13) $$u(n) = G_{p-3}a(n) + (G_{p-3} + G_{p-4})b(n) + G_{p-2}\, c(n) - \lambda_u,$$

<u>where</u> λ_u <u>is independent of</u> n.

To have the theorem hold for all $p \geq 1$ we use the recurrence (11.1) to define G_n for negative values of n. In particular we have

n	-3	-2	-1	0	1	2	3
G_n	-1	1	0	0	1	1	2

In the proof of the theorem the following relations emerge.

$$(14.14)\qquad \begin{cases} \lambda_{va} = \lambda_v + G_p + G_{p-1}, \\ \lambda_{vb} = \lambda_v + G_p, \\ \lambda_{vc} = \lambda_v, \end{cases}$$

where v is the weight of p.

15. Estimate for $a(n)$. Let α be the real root of

$$x^3 - x^2 - x - 1 = 0$$

and let β, γ be the complex roots. It is easily verified that

$$(15.1)\qquad G_{n+1} - \alpha G_n = \frac{\gamma^{n+1} - \beta^{n+1}}{\gamma - \beta} .$$

Making use of (15.1) and the second canonical representation we get

Theorem 15.1. The three sequences

$$(15.2)\qquad a(N) - [\alpha N], \quad b(N) - [\alpha^2 N], \quad c(N) - [\alpha^3 N]$$

are all bounded.

Theorem 15.2. The difference $a(N) - [\alpha N]$ is positive infinitely often, negative infinitely often and 0 infinitely often.

It should be noted that we have been unable to find explicit formulas for the functions $a(N)$, $b(N)$, $c(N)$.

We now define the generating functions

$$(15.3)\qquad \varphi_k(x) = \sum_{n \in C_k} x^k \quad (k = 2, 3, 4, \ldots)$$

and

$$(15.4)\qquad \bar{\varphi}_{3k+1}(x) = \sum_{n \in \bar{C}_{3k+1}} x^k \quad (k = 0, 1, 2, \ldots).$$

Then

$$\overline{\varphi}_{3k+1}(x) = \varphi_{3k+1}(x) \qquad (k = 1, 2, 3, \ldots) \tag{15.5}$$

and

$$\overline{\varphi}_1 = \sum_{k=0}^{\infty} \varphi_{3k+2}(x) + \sum_{k=0}^{\infty} \varphi_{3k+3}(x). \tag{15.6}$$

Also it is evident that

$$\frac{x}{1-x} = \sum_{k=2}^{\infty} \varphi_k(x). \tag{15.7}$$

If we put

$$A(x) = \sum_{n=1}^{\infty} x^{a(n)}, \quad B(x) = \sum_{n=1}^{\infty} x^{b(n)}, \quad C(x) = \sum_{n=1}^{\infty} x^{c(n)}, \tag{15.8}$$

then

$$A(x) + B(x) + C(x) = \frac{x}{1-x} \tag{15.9}$$

and

$$A(x) = \sum_{k=0}^{\infty} \varphi_{3k+3}(x), \tag{15.10}$$

$$B(x) = \sum_{k=0}^{\infty} \varphi_{3k+3}(x), \tag{15.11}$$

$$C(x) = \sum_{k=0}^{\infty} \varphi_{3k+4}(x) = \frac{x}{1-x} - \overline{\varphi}_1(x). \tag{15.12}$$

It follows from

$$\varphi_2(x) = \sum_{n=1}^{\infty} x^{a^2(n)} + \sum_{n=1}^{\infty} x^{ab(n)}$$

and

$$a^2(n) = b(n) - 1, \quad ab(n) = c(n) - 1$$

that

(15.13) $$x\varphi_2(x) = B(x) + C(x).$$

Similarly from

$$\varphi_3(x) = \sum_{n=1}^{\infty} x^{ba(n)} + \sum_{n=1}^{\infty} x^{b^2(n)}$$

we get

(15.14) $$x^2\varphi_3(x) = xA(x) - B(x).$$

We have also

(15.15) $$x^4\varphi_4(x) = x^2B(x) - C(x)$$

and

(15.16) $$x^8\varphi_5(x) = -x\,A(x) + B(x) + (1+x^4)C(x).$$

Generally we have

(15.17) $$x^{S_{k-1}}\varphi_k(x) = p_1(x)A(x) + p_2(x)B(x) + p_3(x)C(x),$$

where

$$S_k = G_1 + G_2 + \cdots + G_k$$

and $p_1(x)$, $p_2(x)$, $p_3(x)$ are polynomials with integral coefficients.

It follows from Theorem 15.1 that

(15.18) $$\sum_{a(k) \le n} 1 \sim \frac{n}{\alpha}, \qquad \sum_{b(k) \le n} 1 \sim \frac{n}{\alpha^2}, \qquad \sum_{c(k) \le n} 1 \sim \frac{n}{\alpha^3}.$$

Making use of (15.18) we get

Theorem 15.3. *The functions* $A(x)$, $B(x)$, $C(x)$ *cannot be continued analytically across the unit circle.*

Theorem 15.4. *The functions* $A(x)$, $B(x)$, $C(x)$ *do not satisfy any relation of the form*

$$f_1(x)A(x) + f_2(x)B(x) + f_3(x)C(x) = 0,$$

where $f_1(x)$, $f_2(x)$, $f_3(x)$ *are polynomials.*

16. Proof of conjectures. Let $\nu_k(N)$ denote the number of numbers $n \in C_k$ such that $n \le N$.

Theorem 16.1. *If* $N \notin C_2$ *then*

$$\nu_2(N) = N - f(N). \tag{16.1}$$

More generally if

$$N \notin C_2 \cup C_3 \cup \cdots \cup C_r$$

then

$$\nu_r(N) = f^{r-2}(N) - f^{r-1}(N) \quad (r = 2, 3, 4, \ldots). \tag{16.2}$$

The following theorem is an immediate corollary.

Theorem 16.2. *We have*

$$\nu_2(G_n) = G_n - G_{n-1} = G_{n-2} + G_{n-3} \quad (n \ge 3). \tag{16.3}$$

More generally

$$\nu_r(G_n) = G_{n-r+2} - G_{n-r+1} = G_{n-r} + G_{n-r-1} \quad (n \ge r + 1). \tag{16.4}$$

The result (16.3) was conjectured by Hoggatt. To prove another conjecture we require the following

Lemma. *Let* $k \ge 1$. *Then there exist integers* A, B *independent of* n *such that*

$$kG_n = \sum_{j=A}^{B} \epsilon_j G_{n+j},$$

where $\epsilon_j = 0$ *or* 1,

$$\epsilon_j\epsilon_{j+1}\epsilon_{j+2} = 0 \quad (j = A, A+1, \ldots, B-2)$$

and the sequence

$$(\epsilon_A, \epsilon_{A+1}, \ldots, \epsilon_B)$$

is independent of n.

We remark that A may be negative.

Applying the lemma to Theorem 16.1 we get

Theorem 16.3. *Let* k, r *be fixed integers*, $k \ge 1$, $r \ge 2$. *Then*

$$\nu_r(kG_n) = k(G_{n-r} + G_{n-r-1}) \tag{16.5}$$

for $n \ge n_0(k,r)$.

It should be observed that results like the above hold for ordinary Fibonacci numbers. Let $\mu_r(N)$ denote the number of numbers $n \in A_k$ such that $n \le N$, where A_k has the same meaning as in the first part of the present paper. Then corresponding to Theorem 16.1 we have

Theorem 16.4. *If* $N \notin A_2$ *then*

$$\mu_2(N) = N - e(N). \tag{16.6}$$

More generally if

$$N \notin A_2 \cup \cdots \cup A_r$$

then

$$\mu_r(N) = e^{r-2}(N) - e^{r-3}(N) \quad (r = 2, 3, 4, \ldots). \tag{16.7}$$

As a corollary we have

Theorem 16.5.

$$\mu_2(F_n) = F_{n-1} \quad (n \ge 3). \tag{16.8}$$

More generally

$$\mu_r(F_n) = F_{n-r} \quad (n > r). \tag{16.9}$$

The lemma continues to hold with F_n replacing G_n. Finally, corresponding to Theorem 16.3 we have

Theorem 16.6. *Let* k, r *be fixed integers*, $k \ge 1$, $r \ge 2$. *Then*

(16.10) $$\mu_r(kF_n) = kG_{n-r}$$

for $n \ge n_0(k,r)$.

IV. PELLIAN REPRESENTATIONS

17. We define the Pellian numbers P_n by means of

$$P_0 = 0, \quad P_1 = 1, \quad P_{n+1} = 2P_n + P_{n-1} \quad (n \ge 1),$$

so that

$$P_2 = 2, \quad P_3 = 5, \quad P_4 = 12, \quad P_5 = 29.$$

Theorem 17.1. *Every positive integer* N *has a unique representation*

(17.1) $$N = \epsilon_1 P_1 + \epsilon_2 P_2 + \epsilon_3 P_3 + \cdots,$$

where $\epsilon_i = 0, 1, 2$ *and in addition*

(17.2) $\epsilon_1 = 0, 1,$

(17.3) *if* $\epsilon_i = 2$ *then* $\epsilon_{i-1} = 0.$

We now define

(17.4) $$e(N) = \epsilon_2 P_1 + \epsilon_3 P_2 + \epsilon_4 P_3 + \cdots,$$

corresponding to $e(N)$ in the Fibonacci case.

We also define a number of additional functions.

(17.5) $a(n) = [\sqrt{2}\, n],$

(17.6) $b(n) = [(2 + \sqrt{2})n] = a(n) + 2n,$

(17.7) $d(n) = [(1 + \sqrt{2})n] = a(n) + n,$

(17.8) $d'(n) = [\tfrac{1}{2}(2 + \sqrt{2})n].$

We recall that if α, β are positive real irrationals such that

$$\frac{1}{\alpha} + \frac{1}{\beta} = 1,$$

then the sets of integers

$$A = \{[\alpha n]\}, \quad B = \{[\beta n]\} \qquad (n = 1, 2, 3, \ldots)$$

constitute a disjoint partition of the positive integers. For brevity we shall say that <u>any</u> two strictly monotone functions A and B are complementary if $A(\underset{\sim}{N})$ and $B(\underset{\sim}{N})$ form a partition of the positive integers.

We define

$$(17.9) \qquad \delta(n) = b(n) + d(n) = 2a(n) + 3n$$

and let ϵ be the function complementary to δ. Since

$$\frac{1}{\sqrt{2}} + \frac{1}{2+\sqrt{2}} = 1, \qquad \frac{1}{1+\sqrt{2}} + \frac{2}{2+\sqrt{2}} = 1,$$

it follows that the $a(n), b(n);\ d(n), d'(n)$ are pairs of complementary functions.

<u>Theorem</u> 17.2. <u>The following relations hold</u>.

(17.10) $\quad \delta(n) = d'b(n), \quad ab(n) = 2d(n), \quad d(n) = a(b(n)-d'(n)),$

(17.11) $\quad ab(n) = a(n) + b(n), \quad db(n) = bd(n) + 1.$

<u>Moreover</u>

(17.12) $\quad e(b(n)) = a(n), \quad e(d(n)) = n, \quad e(\delta(n)) = d(n).$

The following table is easily computed.

n	1	2	3	4	5	6	7	8	9	10	11	12
a	1	2	4	5	7	8	9	11	12	14	15	16
b	3	6	10	13	17	20	23	27	30	34	37	40
d	2	4	7	9	12	14	16	19	21	24	26	28
d′	1	3	5	6	8	10	11	13	15	17	18	20
δ	5	10	17	22	29	34	39	46	51	58	63	68
ε	1	2	3	4	6	7	8	9	11	12	13	14

Let A_k denote the set of integers

$$N = \epsilon_k P_k + \epsilon_{k+1} P_{k+1} + \cdots$$

satisfying the conditions of Theorem 17.1 and in addition $\epsilon_k \neq 0$. Let B_k denote the set of such N with $\epsilon_k = 2$.

Theorem 17.3. The following relations hold.

$$A_1 = d(\underset{\sim}{N}) - 1,$$

$$A_{2k} = d\delta^{k-1}\epsilon(\underset{\sim}{N}) \qquad (k = 1, 2, 3, \ldots),$$

$$A_{2k+1} = \delta^{k}\epsilon(\underset{\sim}{N}) \qquad (k = 1, 2, 3, \ldots),$$

$$B_{2k} = d\delta^{k-1}d(\underset{\sim}{N}) \qquad (k = 1, 2, 3, \ldots),$$

$$B_{2k+1} = \delta^{k}d(\underset{\sim}{N}) \qquad (k = 1, 2, 3, \ldots),$$

$$d(\underset{\sim}{N}) = \bigcup_{k=1}^{\infty} A_{2k}, \quad \delta(\underset{\sim}{N}) = \bigcup_{k=1}^{\infty} A_{2k+1},$$

$$\epsilon(\underset{\sim}{N}) = d(\underset{\sim}{N}) \cup (d(\underset{\sim}{N}) - 1),$$

$$A_{2k} - B_{2k} = d\delta^{k-1}(d(\underset{\sim}{N}) - 1) \qquad (k = 1, 2, 3, \ldots),$$

$$A_{2k+1} - B_{2k+1} = \delta^{k}(d(\underset{\sim}{N}) - 1) \qquad (k = 1, 2, 3, \ldots).$$

The set of numbers $b(\underset{\sim}{N})$ is apparently characterized by the following properties:

(i) the first non-vanishing ϵ_i in the canonical representation (17.1) has an odd subscript;

(ii) the digit

$$\epsilon_1 + \epsilon_2 + \epsilon_3 + \cdots$$

is even.

This suggests various related questions.

V. A SPECIAL SEQUENCE

18. We define a sequence $\{u_n\}$ by means of

(18.1) $$u_0 = 0, \quad u_1 = 1, \quad u_{n+1} = u_n + 2u_{n-1} \quad (n \geq 1).$$

This evidently implies

(18.2) $$u_n = \frac{1}{3}\left(2^n - (-1)^n\right).$$

The first few values of u_n follow.

n	1	2	3	4	5	6	7	8	9	10
u_n	1	1	3	5	11	21	43	85	171	341

It is not difficult to show that the sums

(18.3) $$\sum_{i=2}^{k} \epsilon_i u_i \quad (k = 2, 3, 4, \ldots),$$

where each $\epsilon_i = 0$ or 1, are distinct. The first few numbers (18.3) are

(18.4) 1, 3, 4, 6, 8, 9, 11, 12, 14, 16, 17, 19, 20, ...

Thus there is a sequence of "missing" numbers

(18.5) 2, 7, 10, 13, 18, 23, 28, 31, 34, 39, ...

To identify the sequence (18.5) we first define an array R in the following way. The elements of the first row are denoted by $a(n)$, of the second row by $b(n)$, of the third row by $c(n)$. Put

$$a(1) = 1, \quad b(1) = 3, \quad c(1) = 2.$$

Assume that the first $n-1$ columns of R have been filled. Then $a(n)$ is the smallest integer not already appearing, while

(18.6) $$b(n) = a(n) + 2n$$

and

(18.7) $$c(n) = b(n) - 1.$$

The following table is easily computed.

n	1	2	3	4	5	6	7	8	9	10
a(n)	1	4	5	6	9	12	15	16	17	20
b(n)	3	8	11	14	19	24	29	32	35	40
c(n)	2	7	10	13	18	23	28	31	34	39

Theorem 18.1. The sets $\{a(n)\}$, $\{b(n)\}$, $\{c(n)\}$ constitute a disjoint partition of the positive integers.

Let A_k denote the set of numbers

(18.8) $$N = u_{k_1} + u_{k_2} + \cdots + u_{k_r},$$

where

$$2 \le k = k_1 < k_2 < \cdots < k_r$$

and $r = 1, 2, 3, \ldots$.

Theorem 18.2. We have

(18.9) $$A_{2k+2} = ab^k a(\underset{\sim}{N}) \cup ab^k c(\underset{\sim}{N}) \qquad (k \ge 0),$$

(18.10) $$A_{2k+1} = b^k a(\underset{\sim}{N}) \cup b^k c(\underset{\sim}{N}) \qquad (k \ge 1).$$

Put

(18.11) $$e(N) = u_{k_1-1} + u_{k_2-1} + \cdots + u_{k_r-1},$$

where N is given by (18.8). Then we have

Theorem 18.3. The following relations hold.

(18.12) $$e(a(n)) = n,$$

(18.13) $$e(b(n)) = a(n).$$

As a corollary we state

Theorem 18.4. *Every positive integer* N *can be represented in the form*

$$N = u_{k_1} + u_{k_2} + \cdots + u_{k_r}, \tag{18.14}$$

where

$$1 \le k_1 < k_2 < \cdots < k_r$$

and $r = 1, 2, 3, \ldots$.

The functions a, b, c satisfy various relations. In particular

$$\left\{\begin{array}{l} a^2(n) = b(n) - 2 = a(n) + 2n - 2, \\ ab(n) = ba(n) + 2 = 2a(n) + b(n), \\ ac(n) = ca(n) + 2 = 2a(n) + c(n), \\ cb(n) = bc(n) + 2. \end{array}\right. \tag{18.15}$$

Also if we define

$$d(n) = a(n) + n, \tag{18.16}$$

then

$$\left\{\begin{array}{l} da(n) = 2d(n) - 2, \\ db(n) = 4d(n), \\ dc(n) = 4d(n) - 2. \end{array}\right. \tag{18.17}$$

Comparison of (18.3) with (18.8) yields

Theorem 18.5. *If* N *is a "missing" number then, in the representation (18.14),* $k_1 = 1$, $k_2 = 2$.

Theorem 18.6. *The numbers* $N = a(n)$ *are precisely those for which, in the representation (18.8),* k_1 *is even. Hence in (18.14) we may assume that* k_1 *is odd, in which case the representation is unique.*

Theorem 18.7. *The numbers* c(n) *coincide with the "missing" numbers (18.5).*

We define

(18.18) $$f(N) = \sum_{i=1}^{r} (-1)^{k_i},$$

where N is given by (18.14) with k_1 odd. It is easily verified that

(18.19) $$a(N) - 2N = f(N)$$

and

(18.20) $$a(N) + N = d(N) = \sum_{i=1}^{r} 2^{k_i},$$

by (18.16). It follows from this that (18.14) is closely related to the binary representation.

REFERENCES

1. Problem 5252, Amer. Math. Monthly 71 (1964), 1138; Solution, Amer. Math. Monthly 72 (1965), 1144-1145.
2. L. Carlitz, Fibonacci representations, Fibonacci Quart. 6 (1968), 193-220.
3. L. Carlitz, Fibonacci representations II, Fibonacci Quart. 8 (1970), 113-134.
4. L. Carlitz, V. E. Hoggatt and Richard Scoville, Fibonacci representations, Fibonacci Quart. (to appear).
5. L. Carlitz, V. E. Hoggatt and Richard Scoville, Lucas representations, Fibonacci Quart. (to appear).
6. L. Carlitz, V. E. Hoggatt and Richard Scoville, Fibonacci representations of higher order, Fibonacci Quart. (to appear).
7. L. Carlitz, V. E. Hoggatt and Richard Scoville, Fibonacci representations of higher order II, (in preparation).
8. L. Carlitz, V. E. Hoggatt and Richard Scoville, Pellian representations, (in preparation).
9. L. Carlitz, V. E. Hoggatt and Richart Scoville, A special sequence, (in preparation).

ON THE CLASSIFICATION AND EVALUATION OF SOME PARTITION FUNCTIONS AND THEIR TABLES

L. M. Chawla and John E. Maxfield, Kansas State University

1. Let $p(n)$ denote the number of unrestricted partitions of a positive integer n. For a given positive integer $m \geq 1$, let $p(n,m)$ denote the number of partitions of n of the type

$$(1) \qquad x_1 + \cdots + x_\lambda = n, \quad x_1 \cdots x_\lambda = m, \quad \ldots \quad (1 \leq \lambda \leq n).$$

We say m is <u>admissible</u>, if $p(n,m) \geq 1$.

Let $C_p(n)$ be the number of all admissible values of m, which satisfy (1), for all the partitions $p(n)$ and let C_p denote the set of all these values of m. It is evident that $p(n) = \sum_{m \in C_p} p(n,m)$.

In (1), let $m = \rho n$ and let $t(n)$ denote the number of partitions of n of the type

$$(2) \quad x_1 + \cdots + x_\lambda = n, \quad x_1 \cdots x_\lambda = \rho n, \quad \ldots \quad (1 \leq \lambda \leq n, \quad \rho \geq 1).$$

For a given admissible ρ, let $p(n,\rho n)$ denote the number of partitions of n such that the product of all the summands in every partition is ρn. Let $C_t(n)$ be the number of all admissible values of ρ, which satisfy (2), for all the partitions $t(n)$ and let C_t denote the set of these values of ρ. It is clear that $t(n) = \sum_{\rho \in C_t} p(n,\rho n)$.

Finally in (1), let $m = n^\rho$ and let $r(n)$ denote the number of partitions of n of the type

$$(3) \quad x_1 + \cdots + x_\lambda = n, \quad x_1 \cdots x_\lambda = n^\rho, \quad \ldots \quad (1 \leq \lambda \leq n, \quad \rho \geq 1).$$

Let $C_r(n)$ and C_r and $p(n,n^\rho)$ be defined as above. Then $r(n) = \sum_{\rho \in C_r} p(n,n^\rho)$.

We need hardly point out that the partitions $p(n)$, considered as a set, are classified into $C_p(n)$ classes $p(n,m)$. Similarly, the partitions $t(n)$ and $r(n)$ are classified respectively into $C_t(n)$ and $C_r(n)$ classes $p(n,\rho n)$ and $p(n,n^\rho)$.

In the present paper, we determine C_p, $C_p(n)$, $p(n,m)$; C_t, $C_t(n)$, $p(n,\rho n)$; C_r, $C_r(n)$, $p(n,n^\rho)$ by means and in terms of a related pair of functions $f(n)$ and $q(n)$ studied in [1] and recalled below. We also tabulate C_t, $C_t(n)$, $t(n)$ up to $n = 27$ (see Table 1) and C_r, $C_r(n)$, $r(n)$ up to $n = 50$ (see Table 2).

Let $n = p_1^{a_1}\cdots p_r^{a_r}$, define $f(n) = a_1p_1 + \cdots + a_rp_r$, where $f(1) = 0$. Then $f(n) \le n$ for all n and the equality holds only for $n = 4$ or when n is a prime [1, Lemmas 1.1 and 1.4]. For a given $n \ge 2$, $q(n)$ is defined to be the set of integers x such that $f(x) = n$ and we set $q(0) = 1$ and $q(1) = 0$. Then $q(n)$ is the number of partitions of n over the set of primes. The tables of $q(n)$ up to $n = 150$ were constructed in [2]. We are now in a position to prove the following:

Lemma 1.1.

(i) $p(n,m) \ge 1$ if and only if $f(m) \le n$,

(ii) $p(n,m) = 0$ if $f(m) > n$,

(iii) $p(n,m) = p(m,m)$ if $1 \le m \le n$,

(iv) $p(n,m) = 0$ if m is a prime $> n$,

(v) $p(n,m) = 1$ if m is a prime $\le n$.

Let $x_1 + \cdots + x_\lambda = n$, $x_1\cdots x_\lambda = m$. Hence

$$f(m) = f(x_1\cdots x_\lambda) = f(x_1) + \cdots + f(x_\lambda) \le x_1 + \cdots + x_\lambda = n.$$

Conversely if $f(m) \le n$, then if $m = q_1^{b_1}\cdots q_s^{b_s}$, we have $f(m) = b_1q_1 + \cdots + b_sq_s \le n$. Hence n can be partitioned at least in one way as $(q_1, \ldots, q_1, q_2, \ldots, q_2, \ldots, q_s, \ldots, q_s, 1, \ldots, 1)$, where the product of all the summands is m and their sum is n. this proves (i).

If $f(m) > n$, we cannot have $x_1 + \cdots + x_\lambda = n$, $x_1\cdots x_\lambda = m$, since this would imply

$$f(m) = f(x_1\cdots x_\lambda) = f(x_1) + \cdots + f(x_\lambda) \le x_1 + \cdots + x_\lambda = n.$$

Hence $p(n,m) = 0$. This proves (ii).

To prove (iii), we first note that $f(m) \le m$ and hence $f(m) \le n$, since $1 \le m \le n$. Thus $p(n,m) \ge 1$, by (i). Also $p(m,m) \ge 1$ by (i). We next note that in the present case every partition in $p(n,m)$ is of the type $(x_1, \ldots, x_\alpha, 1, 1, 1, \ldots, 1)$, where $x_1 + \cdots + x_\alpha + 1 + \cdots + 1 = n$, $x_1 \cdots x_\alpha = m$, and the number μ of unities in the partition satisfies $n - m \le \mu \le n - f(m)$. Hence from each partition in $p(n,m)$, we get a unique partition in $p(m,m)$ by deleting the necessary number of unities and conversely from each partition in $p(m,m)$ we can get a unique partition in $p(n,m)$ by adding the necessary number of unities. This proves (iii).

(iv) follows from (ii), since if m is a prime $> n$ then $f(m) = m > n$.

(v) follows from (iii), since $p(m,m) = 1$, if m is a prime.

Theorem 1.2.

(i) $C_p(n) = q(0) + q(1) + \cdots + q(n)$,

(ii) $C_p = \{x \in Z \mid f(x) = j,\ j = 0, 1, \ldots, n\}$ where Z is the set of all positive integers.

Alternatively, C_p is the set of all integers m obtained by multiplying the summands in each of the partitions $q(j)$, $0 \le j \le n$, with the convention that $q(0)$ gives $m = 1$ and $q(1)$ gives no value of m.

We know from Lemma 1.1(i) that n can be partitioned as $x_1 + \cdots + x_\lambda = n$, $x_1 \cdots x_\lambda = m$, if and only if $f(m) \le n$. Hence by the definition of $q(n)$, the number of all the admissible values of m is given by $C_p(n) = q(0) + q(1) + \cdots + q(n)$. The actual values of m are then just the positive integers x which satisfy $f(x) = j$, $0 \le j \le n$, and are therefore just the integers obtained by multiplying the summands in each of the partitions $q(j)$, $0 \le j \le n$, with the convention as stated in the theorem. This completes the proof.

From equations (2) and Lemma 1.1(i), it follows that n can be partitioned as $x_1 + \cdots + x_\lambda = n$, $x_1 \cdots x_\lambda = \rho n$, $1 \le \lambda \le n$, $\rho \ge 1$, if and only if $f(\rho n) = f(\rho) + f(n) \le n$ or if and only if $f(\rho) \le n - f(n) \equiv d$. Hence we have the following analogue of Theorem 1.2 for the partitions $t(n)$.

Theorem 1.3.

(i) $C_t(n) = q(0) + q(1) + \cdots + q(d)$, $d = n - f(n)$,

(ii) $C_t = \{x \in Z \mid f(x) = j,\ j = 0, 1, 2, \ldots, d\}$.

Alternatively, the set C_t of positive integral values of ρ which satisfy (2) is just the set of integers obtained by multiplying the summands in each of the partitions $q(j)$, $j = 0, 1, 2, \ldots, d$, with the convention that $q(0)$ gives $\rho = 1$ and $q(1)$ gives no value of ρ.

From equations (3) and Lemma 1.1(i), it follows that n can be partitioned as $x_1 + \cdots + x_\lambda = n$, $x_1 \cdots x_\lambda = n^\rho$, $1 \le \lambda \le n$, $\rho \ge 1$, if and only if $f(n^\rho) = \rho f(n) \le n$ or if and only if $\rho \le \left[\frac{n}{f(n)}\right]$. Hence we have the following analogue of Theorem 1.2 for the partitions $r(n)$.

Theorem 1.4.

(i) $C_r(n) = \left[\frac{n}{f(n)}\right]$,

(ii) $C_r = \left\{1, 2, 3, \ldots, \left[\frac{n}{f(n)}\right]\right\}$.

We close this section with the observation that if in the equations $x_1 + \cdots + x_\lambda = n$, $x_1 \cdots x_\lambda = m$, $f(m) \le n$ we restrict the values of m by $f(m) = n$, it follows that the resulting partitions of n are just the partitions $q(n)$ of n over the set of primes. In this case $C_q(n) = q(n)$ and thus each such partition is in a class by itself. Further, the set C_q for $q(n)$ is the set of integers obtained by multiplying the summands in each of the partitions $q(n)$.

2. In this section we shall evaluate the partition functions $p(n,m)$, $p(n,\rho n)$ and $p(n,n^\rho)$ and thereby the partition functions $p(n)$, $t(n)$ and $r(n)$.

Let $d_1, \ldots, d_\ell$ denote all the divisors of m which satisfy $1 < d_i < n$ if $m > n$, and $1 < d_i \le n$ if $m = n$. The case $m < n$ need not be considered separately, since in that case we have $p(n,m) = p(m,m)$ by Lemma 1.1(iii).

Let N_{d_i} denote the number of partitions in $p(n,m)$ which contain d_i as a summand at least once. Let N_{d_i,d_j}, $i \neq j$, denote the number of partitions in $p(n,m)$, which contain d_i, d_j as summands at least once. Define $N_{d_i,d_j,d_k,\ldots}$ for every possible non-empty subset $\{d_i, d_j, d_k, \ldots\}$ of $\{d_1, d_2, \ldots, d_\ell\}$. We then have:

<u>Theorem</u> 2.1. $$p(n) = \sum_{m \in C_p} p(n,m),$$

where

$$p(n,m) = \sum N_{d_i} - \sum N_{d_i,d_j} + \sum N_{d_i,d_j,d_k} - \cdots .$$

Further,

(i) $N_{d_i} = p(n-d_i, m/d_i)$,

(ii) $N_{d_i} = p(n-d_i, m/d_i) = p(m/d_i, m/d_i)$ if $m/d_k \le n - d_i$,

(iii) $N_{d_i} = 0$ if $f(m/d_i) > n - d_i$,

(iv) $N_{d_i,d_j} = p(n-d_i-d_j, m/d_i d_j)$ if $d_i + d_j < n$ and $d_i d_j | m$,

(v) $N_{d_i,d_j} = p(n-d_i-d_j, m/d_i d_j) = p(m/d_i d_j, m/d_i d_j)$ if $d_i + d_j + m/d_i d_j \le n$ and $d_i d_j | m$,

(vi) $N_{d_i,d_j} = 0$ if $d_i d_j \nmid m$, or if $d_i + d_j > n$, or if $d_i d_j | m$ and $d_i + d_j < n$ but $f(m/d_i d_j) > n - d_i - d_j$.

Results similar to (iv)-(vi) hold for the remaining $N_{d_i,d_j,d_k,\ldots}$.

Since every partition in $p(n,m)$ is counted by one or the other of the partitions N_{d_i}, N_{d_i,d_j}, $N_{d_i,d_j,d_k}, \ldots$, the principle of cross-classification implies

$$0 = p(n,m) - \sum N_{d_i} + \sum N_{d_i,d_j} - \sum N_{d_i,d_j,d_k} + \cdots .$$

Hence

$$p(n,m) = \sum N_{d_i} - \sum N_{d_i,d_j} + \sum N_{d_i,d_j,d_k} - \cdots .$$

Also we have $p(n) = \sum_{m \in C_p} p(n,m)$ as noted in the beginning. This proves the first part of the theorem.

To prove (i), we note that $p(n-d_i, m/d_i)$ just counts all the partitions in $p(n,m)$, which have d_i as a summand at least once. Hence $N_{d_i} = p(n-d_i, m/d_i)$.

(ii) follows from (i) combined with Lemma 1.1(iii). Next (iii) follows from Lemma 1.1(ii). The remaining (iv)-(vi) can be proved on the same lines as (i)-(iii).

<u>Corollary</u>.

$$p(m,m) = \sum p(m/d_i, m/d_i) - \sum p(m/d_id_j, m/d_id_j)$$

$$+ \sum p(m/d_id_jd_k, m/d_id_jd_k) - \cdots .$$

This follows immediately from

$$p(m,m) = \sum N_{d_i} - \sum N_{d_i,d_j} + \sum N_{d_i,d_j,d_k} - \cdots$$

and (ii) and (v) in Theorem 2.1.

With the same notations as in Theorem 2.1, if $d_1, \ldots, d_\ell$ are all the divisors of ρn (or of n^ρ) such that $1 < d_i < \rho n$ (or n^ρ) if $\rho > 1$, and $1 < d_i \leq n$ if $\rho = 1$, then for the partitions $p(n, \rho n)$, $t(n)$, and $p(n, n^\rho)$ and $r(n)$, we have the following analogues of Theorem 2.1.

<u>Theorem</u> 2.2. For a prime n, $t(n) = 1$. For a composite n,

$$t(n) = \sum_{\rho \in C_t} p(n, \rho n),$$

where

$$p(n,\rho n) = \sum N_{d_i} - \sum N_{d_i,d_j} + \sum N_{d_i,d_j,d_k} - \cdots .$$

Furthermore, the results (i)-(vi) hold when m is replaced by ρn in Theorem 2.1.

Theorem 2.3. For a prime n, $r(n) = 1$. For a composite n,

$$r(n) = \sum_{\rho \in C_r} p(n,n^\rho),$$

where

$$p(n,n^\rho) = N_{d_i} - \sum N_{d_i,d_j} + \sum N_{d_i,d_j,d_k} - \cdots .$$

Furthermore, the results (i)-(vi) hold when m is replaced by n^ρ in Theorem 2.1.

3. For the tabulation of C_t, $C_t(n)$, $t(n)$, and C_r, $C_r(n)$ and $r(n)$, we first observe that since n^ρ can be written as sn, where $s = n^{\rho-1}$, it follows that $r(n) \le t(n)$ and further $r(n) = p(n,n) + p(n,n^2) + \cdots + p(n,n^\rho)$, and thus each of the classes of $r(n)$ is a class in $t(n)$.

For the tabulation of C_t, $C_t(n)$, we used Theorem 1.3 above and the tables of $q(n)$ given in [2]. For the tabulation of C_r, $C_r(n)$ we used Theorem 1.4 above.

To tabulate $t(n) = \sum_{\rho \in C_t} p(n,\rho n)$ up to $n = 27$, we proceed as follows. We first determine C_t as indicated above. For a given ρ, let $\rho n = q_1^{b_1} \cdots q_s^{b_s}$. Since $f(\rho n) = b_i q_i + \cdots + b_s q_s \le n$, we have a partition $(q_1, \ldots, q_1, q_2, \ldots, q_2, \ldots, q_s, \ldots, q_s, 1, \ldots, 1)$ of n of the requisite type in $p(n,\rho n)$. All the remaining partitions in $p(n,\rho n)$ are then built up by the admissible multiplications of the summands in the above partition and by checking every step by the use of Theorem 2.2 and the Corollary of Theorem 2.1. Of course, one can also evaluate $p(n,\rho n)$ by summing up the coefficients of all such x^n in

$$\prod(1-x)^{-1}(1-x^{d_1})^{-1}\cdots(1-x^{d_\ell})^{-1},$$

which are of the type $x^n = x^{\alpha_1+\cdots+\alpha_\lambda}$, where $\alpha_1\cdots\alpha_\lambda = \rho n$.

Examples.

(i) Let $n = 15$. Then $d = 15 - f(15) = 15 - 8 = 7$. Hence $C_t(15) = q(0) + q(1) + \cdots + q(7) = 1 + 0 + 1 + 1 + 1 + 2 + 2 + 3 = 11$. Hence $C_t = \{1;\ 2;\ 3;\ 4;\ 5;\ 6;\ 2^3,\ 3^2;\ 7;\ 2\cdot 5,\ 2^2\cdot 3\}$.

$p(15,15) = 2$, $p(15,2\cdot 15) = 3$, $p(15,3\cdot 15) = 2$, $p(15,2^2\cdot 15) = 4$, $p(15,5\cdot 15) = 1$, $p(15,2\cdot 3\cdot 15) = 2$, $p(15,7\cdot 15) = 1$, $p(15,2^3\cdot 15) = 4$, $p(15,3^2\cdot 15) = 1$, $p(15,2\cdot 5\cdot 15) = 1$, $p(15,2^2\cdot 3\cdot 15) = 2$. Hence $t(15) = 23$.

For $n = 15$, we have $[n/f(n)] = [15/8] = 1$. Hence $C_r(15) = 1$, $C_r = \{1\}$. Thus $r(15) = p(15,15) = 2$.

(ii) Let $n = 16$, then $d = 16 - f(16) = 16 - 8 = 8$. Hence $C_t(16) = q(0) + q(1) + \cdots + q(8) = 1 + 0 + 1 + 1 + 1 + 2 + 2 + 3 + 3 = 14$. $C_t = \{1;\ 2;\ 3;\ 2^2;\ 5,\ 2\cdot 3;\ 2^3,\ 3^2;\ 7;\ 2\cdot 5,\ 2^2\cdot 3;\ 3\cdot 5,\ 2^4,\ 2\cdot 3^2\}$.

Now $p(16,16) = 5$, $p(16,2\cdot 16) = 5$, $p(16,3\cdot 16) = 8$, $p(16,2^2\cdot 16) = 7$, $p(16,5\cdot 16) = 6$, $p(16,2\cdot 3\cdot 16) = 9$, $p(16,7\cdot 16) = 3$, $p(16,2^3\cdot 16) = 7$, $p(16,3^2\cdot 16) = 8$, $p(16,2\cdot 5\cdot 16) = 3$, $p(16,2^2\cdot 3\cdot 16) = 8$, $p(16,3\cdot 5\cdot 16) = 3$, $p(16,2^4\cdot 16) = 5$, $p(16,2\cdot 3^2\cdot 16) = 3$. Hence $t(16) = 80$.

Further, for $n = 16 = 2^4$, we have $[n/f(n)] = [16/8] = 2$. Hence $C_r(16) = 2$, $C_r = \{1,\ 2\}$. Thus $r(16) = p(16,16) + p(16,16^2)$ $= 5 + 5 = 10$.

Table 1

n	$C_t(n)$	C_t	t(n)
1	1	1	1
2	1	1	1
3	1	1	1
4	1	1	2
5	1	1	1
6	1	1	2
7	1	1	1
8	2	1,2	6
9	3	1,2,3	5
10	3	1,2,3	5
11	1	1	1
12	6	1,2,3,4,5,6	22
13	1	1	1
14	6	1,2,3,4,5,6	11
15	11	1,2,3,4,5,6,7,8,9,10,12	23
16	14	1,2,3,4,5,6,7,8,9,10,12,15,16,18	80
17	1	1	1
18	23	1,2,3,4,5,6,7,8,9,10,11,12,14,15,16,18, 20,21,24,25,27,30,32	112
19	1	1	1
20	29	1,2,3,4,5,6,7,8,9,10,11,12,14,15,16,18 20,21,24,25,27,28,30,32,36,40,45,48,54	150
21	29	same as 20	85
22	18	1,2,3,4,5,6,7,8,9,10,12,14,15,16,18,20, 24,27	45
23	1	1	1
24	67	1,2,3,4,5,6,7,8,9,10,11,12,13,14,15,16, 18,20,21,22,24,25,26,27,28,30,32,33,35, 36,40,42,44,45,48,49,50,54,56,60,63,64, 70,72,75,80,81,82,90,96,100,105,108,112, 120,125,126,128,135,144,150,160,162,180, 192,216,243	737
25	67	same as 24	226
26	29	same as 20	84
27	115	1,2,3,4,5,6,7,8,9,10,11,12,13,14,15,16, 17,18,20,21,22,24,25,26,27,28,30,32,33, 35,36,39,40,42,44,45,48,49,50,52,54,55, 56,60,63,64,65,66,70,72,75,77,78,80,81, 84,88,90,96,98,99,100,105,108,110,112, 120,125,126,128,132,140,144,147,150,160, 162,168,180,185,189,192,196,200,210,216, 224,225,240,243,250,252,256,270,280,288, 300,315,320,324,336,360,375,378,384,400, 405,432,450,480,486,540,576,648,729	803

Table 2

n	$C_r(n)$	C_n	$r(n)$	n	$C_r(n)$	C_n	$r(n)$
1	1	1	1	26	1	1	2
2	1	1	1	27	3	1,2,3	8
3	1	1	1	28	2	1,2	10
4	1	1	2	29	1	1	1
5	1	1	1	30	3	1,2,3	24
6	1	1	2	31	1	1	1
7	1	1	1	32	1	1,2,3	43
8	1	1	3	33	2	1,2	4
9	1	1	2	34	1	1	2
10	1	1	2	35	2	1,2	3
11	1	1	1	36	3	1,2,3	106
12	1	1	4	37	1	1	1
13	1	1	1	38	1	1	2
14	1	1	2	39	2	1,2	4
15	1	1	2	40	3	1,2,3	66
16	2	1,2	10	41	1	1	1
17	1	1	1	42	3	1,2,3	43
18	2	1,2	8	43	1	1	1
19	1	1	1	44	2	1,2	12
20	2	1,2	8	45	4	1,2,3,4	22
21	2	1,2	3	46	1	1	2
22	1	1	2	47	1	1	1
23	1	1	1	48	4	1,2,3,4	465
24	2	1,2	34	49	3	1,2,3	4
25	2	1,2	3	50	4	1,2,3,4	27

REFERENCES

1. L. M. Chawla, On a pair of arithmetic functions, J. Natur. Sci. and Math. 8 (1968), 263-269.

2. L. M. Chawla and S. A. Shad, On a trio-set of partition functions and their tables, J. Natur. Sci. and Math. 9 (1969), 87-96.

TWO OSCILLATION THEOREMS*

Harold G. Diamond, University of Illinois

Ingham established in [7] a method of estimating from below the oscillation of a real-valued function in terms of the singularities of its Laplace transform. This technique and a clever searching procedure with an electronic computer have led to the disproof of various conjectures in number theory (cf. Haselgrove [6], Lehman [9]). In this note we shall give some theorems which are analogous to Ingham's, but are established in a rather different way.

Let f be a real-valued measurable function on $[0,\infty)$ and let $F(s) = \int_0^\infty e^{-su} f(u)\,du$. We assume that the integral converges for $\operatorname{Re} s > 0$ and that F can be continued as a meromorphic function to a region of the complex plane which includes the imaginary axis. Assume further that there are poles of F on the imaginary axis and that they are all simple. Let

$$T = \{t_1, t_2, \ldots\} = \{t > 0 : ti \text{ is a pole of } F\},$$

a finite or countable set. Let a_n be the residue of F at it_n and let a_0 be the residue of F at 0 ($a_0 = 0$ if F is regular at 0).

We say that a finite subset

$$T_1 = \{t_{i_1}, t_{i_2}, \ldots, t_{i_J}\} \subset T$$

is a <u>weakly independent subset of</u> T <u>of order</u> N provided that $\sum_{j=1}^{J} n_j t_{i_j} \in T$ for some integers n_j with $|n_j| \le N$ implies that $\sum_{j=1}^{J} |n_j| = 1$. That is, the only way that a sum $\sum n_j t_{i_j}$ with $|n_j| \le N$ can represent an element of T is that one $n_j = 1$ and all others be zero.

*This research was partially supported by NSF Grant GP 21335.

We formulate our oscillation theorems in terms of the notion of weak independence. Note that if T is linearly independent over the integers, then any finite subset satisfies a weak independence condition of any order. It is not hard to establish the equivalence of our two theorems. We shall prove the second theorem and indicate how one deduces the first.

Theorem 1. Suppose there exist a finite collection of indices $\mathcal{J}$ and a positive integer N such that $\{t_j \in T: j \in \mathcal{J}\}$ is a weakly independent subset of T of order N. Then

$$\lim_{X\to\infty} \operatorname{ess\,sup}_{u \geq X} f(u) \geq a_o + \frac{2N}{N+1} \sum_{j \in \mathcal{J}} |a_j|,$$

$$\lim_{X\to\infty} \operatorname{ess\,inf}_{u \geq X} f(u) \leq a_o - \frac{2N}{N+1} \sum_{j \in \mathcal{J}} |a_j|.$$

Theorem 2. In addition to the hypotheses of Theorem 1, assume that $\frac{2N}{N+1} \sum_{j \in \mathcal{J}} |a_j| > |a_o|$. Then there exists no real number X such that f (or some L^1 equivalent) is of one sign on $[X, \infty)$.

Proof of Theorem 2. The case in which $a_o = 0$, i.e., F regular at 0, follows immediately from Landau's Theorem on Mellin transforms of nonnegative functions [8; Satz 454]. (It could also be handled directly.) We henceforth assume for the proof of Theorem 2 that $a_o > 0$. Let K_N be the N-th Fejér kernel

$$K_N(x) = \sum_{-N}^{N} \left(1 - \frac{|n|}{N+1}\right) e^{inx} = \frac{1}{N+1} \left|\sum_{0}^{N} e^{inx}\right|^2 \geq 0.$$

Let J be the cardinality of $\mathcal{J}$ and let $x_1, x_2, \ldots, x_J$ be a sequence of real numbers. Form the product

$$0 \leq \prod_{j=1}^{J} K_N(x_j) = \sum_{n_1=-N}^{N} \cdots \sum_{n_J=-N}^{N} \prod_{j=1}^{J} \left\{\left(1 - \frac{|n_j|}{N+1}\right) e^{in_j x_j}\right\}$$

$$= 1 + 2\left(1 - \frac{1}{N+1}\right) \sum_{j=1}^{J} \operatorname{Re} e^{ix_j} + S,$$

where S is a real-valued trigonometric polynomial each of whose terms involves at least two x_j's (possibly the same x_j repeated). For notational simplicity we assume that $\mathcal{J} = \{1, 2, \ldots, J\}$. Let $c_1, \ldots, c_J$ be real numbers such that $\arg a_j e^{ic_j} \equiv \pi \pmod{2\pi}$, and set $x_j = -ut_j + c_j$, where $t_j \in T$ and u is an integration variable to be specified.

We assume that the conclusion of Theorem 2 is false, and there exists an X such that f is of one sign on $[X, \infty)$. For $\sigma > 0$ we form

$$\left\{\int_{u=0}^{X} + \int_{X}^{\infty}\right\}\left\{f(u)e^{-\sigma u} \prod_{j=1}^{J} K_N(-ut_j + c_j)\,du\right\}$$

$$= F(\sigma) + \frac{2N}{N+1} \sum_{j=1}^{J} \operatorname{Re}\,\{F(\sigma+it_j)e^{ic_j}\} + \sum e_n F(\sigma+i\tau_n),$$

with suitable e_n's and τ_n's. By the weak independence condition no τ_n lies in T. Now let $\sigma \to 0+$ and note that $\int_{u=0}^{X} = O(1)$, $\sum e_n F(\sigma+i\tau_n) = O(1)$, $\int_{u=X}^{\infty} \geq 0$, and $F(\sigma+it_j)e^{ic_j} \sim -|a_j|\sigma^{-1}$. We conclude that

$$a_0 - \frac{2N}{N+1} \sum_{j=1}^{J} |a_j| \geq 0,$$

in contradiction to one of the hypotheses. Thus Theorem 2 is true.

We sketch the proof of Theorem 1 for the lower bound. Let ϵ be a small positive number and take

$$a_0' = a_0 - \frac{2N}{N+1} \sum_{j \in \mathcal{J}} |a_j| + \epsilon.$$

Set $g(u) = f(u) - a_0'$ and apply Theorem 2 to show that there exist arbitrarily large values of X such that $\operatorname{ess\,inf}_{u \geq X} g(u) < 0$. Changing the last relation to one involving f and letting $\epsilon \to 0+$, we obtain the second estimate of Theorem 1.

Remarks. 1. It is clear from the proof that the meromorphy requirement in both theorems is unnecessarily strong and can be replaced by weaker but messier conditions.

2. The idea of using a product of Fejér kernels plus an independence condition to estimate the values of a function dates back at least to Bohr and Jessen [2].

3. Theorems 1 and 2 yield new proofs of some theorems and offer the possibility of improving some known estimates. In particular, we obtain another proof of the "variant form of Ingham's theorem" of Bateman et al. [1] and can improve estimates of Grosswald [4], [5] and Saffari [10], [11], [12].

4. To obtain better numerical estimates in specific cases, one should alter the proof of Theorem 2 as follows. First, in place of the N-th Fejér kernel one should use the nonnegative trigonometric polynomial

$$P_N^*(t) = \left|\sum_{j=0}^{N} \left(\sin\frac{[j+1]\pi}{N+2}\right)e^{ijt}\right|^2 \left(\sum_{j=0}^{N} \sin^2 \frac{[j-1]\pi}{N+2}\right)^{-2}$$

$$= 1 + \left(2\cos\frac{\pi}{N+2}\right)\cos t + \cdots .$$

It was proved by Fejér [3; page 79] that, for each N, this polynomial has the largest value of λ_1 among all nonnegative trigonometric polynomials of the form

$$1 + \lambda_1 \cos t + \cdots + \lambda_N \cos Nt.$$

Second, in place of the product of J nonnegative trigonometric polynomials each of degree N, one should multiply together the polynomials $P_{N_j}^*(x_j)$, where the N_j are judiciously chosen and are generally not all equal.

With these changes Theorem 1 can be stated as follows: Suppose that $N_1, \ldots, N_J$ are nonnegative integers such that

$$\sum_{j=1}^{J} n_j t_j \in T, \quad |n_j| \le N_j \Rightarrow \sum_{j=1}^{J} |n_j| = 1.$$

Then

$$\lim_{X\to\infty} \operatorname*{ess\,sup}_{u \ge X} f(u) \ge a_o + 2\sum_{j=1}^{J} \left(\cos\frac{\pi}{N_j+2}\right)|a_j|$$

and

$$\lim_{X\to\infty} \operatorname*{ess\,inf}_{u \ge X} f(u) \le a_o - 2\sum_{j=1}^{J} \left(\cos\frac{\pi}{N_j+2}\right)|a_j|.$$

Also, we can replace the hypotheses of Theorem 2 by the above assumption and the inequality

$$2\sum_{j=1}^{J} \left(\cos\frac{\pi}{N_j+2}\right)|a_j| > |a_o|.$$

The conclusion of Theorem 2 then follows.

5. The present theorems appear to be of no use in disproving Mertens' conjecture that $\left|\sum_{n \le x} \mu(n)\right| \le \sqrt{x}$, $x \ge 1$, μ = Möbius' function. The number of weak independence relations to be checked is far beyond the capacity of present day computers. We investigated the applicability of the present method to the Pólya conjecture that $\sum_{n \le x} \lambda(n) \le 0$, $x \ge 2$, λ = Liouville's function. This was shown to be false by Haselgrove [6] using Ingham's result and computing. Using our method, improved as described in the preceding remark, and a table contained in [6], we showed how one could hope to obtain the required weak independence by checking about 10^{12} sums, each involving at most 12 summands. Unfortunately, it would require several weeks of computer time to carry out such a project. This example suggests that Ingham's theorem and "educated" trials on a computer is more practical that our method for numerical work.

I wish to thank Professors Paul T. Bateman and Emil Grosswald for a number of helpful discussions and suggestions.

REFERENCES

1. P. T. Bateman, J. W. Brown, R. S. Hall, K. E. Kloss, and R. M. Stemmler, Linear relations connecting the imaginary parts of the zeros of the zeta function, to appear in Proceedings of the Atlas Symposium on Computers in Number Theory.

2. H. Bohr and B. Jessen, One more proof of Kronecker's theorem, J. London Math. Soc. 7 (1932), 274-275.

3. L. Fejér, Über trigonometrische Polynome, J. Reine Angew. Math. 146 (1916), 53-82.

4. E. Grosswald, On some generalizations of theorems of Landau and Pólya, Israel J. Math. 3 (1965), 211-220.

5. E. Grosswald, Oscillation theorems of arithmetical functions, Trans. Amer. Math. Soc. 126 (1967), 1-28.

6. C. B. Haselgrove, A disproof of a conjecture of Pólya, Mathematika 5 (1958), 141-145.

7. A. E. Ingham, On two conjectures in number theory, Amer. J. Math. 64 (1942), 313-319.

8. E. Landau, Vorlesungen über Zahlentheorie, Band 2, S. Hirzel, Leipzig, 1927.

9. R. S. Lehman, On Liouville's function, Math. Comp. 14 (1960), 311-320.

10. B. Saffari, Sur les oscillations des fonctions sommatoires des fonctions de Möbius et de Liouville, C. R. Acad. Sci. Paris 271 (1970), 578-580.

11. B. Saffari, Sur la fausetté de la conjecture de Mertens, C. R. Acad. Sci. Paris 271 (1970), 1097-1100.

12. B. Saffari, Ω-théorèmes sur le terme résiduel dans le loi de répartition des entiers non divisibles par une puissance $r^{\text{ième}}$, $r > 1$, C. R. Acad. Sci. Paris 272 (1971), 95-97.

ON THE ITERATES OF SOME ARITHMETIC FUNCTIONS

P. Erdös, Hungarian Academy of Sciences
M. V. Subbarao, University of Alberta

1. Introduction. For any arithmetic function $f(n)$, we denote its iterates as follows:

$$f_1(n) = f(n); \quad f_k(n) = f_1[f_{k-1}(n)] \quad (k > 1).$$

Let $\sigma(n)$ and $\sigma^*(n)$ denote, respectively, the sum of the divisors of n, and the sum of its unitary divisors, where we recall that d is called a unitary divisor of n if $(d,n/d) = 1$. Makowski and Schinzel [3] proved that

$$\liminf \frac{\sigma_2(n)}{n} = 1,$$

and conjectured that

$$\liminf \frac{\sigma_k(n)}{n} < \infty \quad \text{for every } k.$$

This is not proved even for $k = 3$. On the other hand, Erdös [2] stated that if we neglect a sequence of density zero, then

$$\frac{\sigma_k(n)}{\sigma_{k-1}(n)} = (1 + o(1))k\, e^{\gamma} \log\log\log n.$$

This implies, in particular, that

$$\frac{\sigma_2(n)}{\sigma_1(n)} \to \infty$$

on a set of density unity.

In contrast to this, we show here the following result.

Theorem 1.

$$\frac{\sigma_2^*(n)}{\sigma_1^*(n)} \to 1 \quad \text{on a set of density unity.}$$

2. Some lemmas. The proof makes use of the following lemmas. Throughout what follows, h, q, r, r_1, r_2 represent primes, and ϵ, η small positive numbers. Almost all $n < x$ will mean: all but $o(x)$ integers $n \le x$.

Lemma 1. For almost all $n < x$, every $p < (\log\log x)^{1-\epsilon}$ satisfies $p^2 | \sigma^*(n)$.

Lemma 2. For almost all $n < x$ and for any given η, we have

$$\sum_{\substack{p|\sigma^*(n) \\ p > (\log\log x)^{1+\epsilon}}} \frac{1}{p} < \eta,$$

where $\epsilon = \epsilon(\eta) > 0$ is sufficiently small.

Lemma 3. For almost all $n < x$ and all $p < t$ (t fixed but arbitrary),

$$p^\alpha | \sigma^*(n)$$

for every fixed α.

We only outline the proofs of the lemmas and the theorem.

Proof of Lemma 1. For a given $p < (\log\log x)^{1-\epsilon}$ for which $p|\sigma_2^*(n)$, $n < x$, it is enough if we show that there are at least two primes r_1, r_2 such that

$$r_1 \equiv r_2 \equiv -1 \pmod p,$$

and

$$r_1|n, \quad r_1^2|n, \quad r_2|n, \quad r_2^2|n.$$

For this purpose we use the Page-Walfisz-Siegel formula for primes in arithmetic progression (Pracher [6], p. 320) which states that if $\pi(a,d,y)$ denotes the number of primes $\equiv a \pmod d$ and $\le y$, then for $(a,d) = 1$,

$$\pi(a,d,y) = (1 + o(1))\frac{y}{\varphi(d)\log y}$$

uniformly in a and d for $d < (\log y)^t$ for every fixed t. Hence, for primes r such that $r|n$, $r \equiv -1 \pmod p$, we have

$$\sum_{\substack{r \equiv -1 \pmod p \\ \log\log x < r < x}} \frac{1}{r} > c(\log\log x)^{\epsilon}.$$

Hence we easily obtain by the sieve of Brun or Selberg that the number of integers $n < x$ which are divisible by just one prime is less than $x \exp(-c(\log\log x)^{\epsilon})$. There are fewer than $(\log\log x)^{1-\epsilon}$ primes $< (\log\log x)^{1-\epsilon}$, and $(\log\log x)^{1-\epsilon} x \exp(-c(\log\log x)^{\epsilon}) = o(x)$, and the number of integers which are divisible by the square of a prime $> \log\log x$ is $o\left(\frac{x}{\log\log x}\right)$. Thus these numbers can be ignored. Thus Lemma 1 is proved.

Proof of Lemma 2. We consider the sum

$$S = \sum_{n=1}^{x} \sum_{\substack{p \mid \sigma^*(n) \\ p > (\log\log x)^{1+\epsilon}}} \frac{1}{p}.$$

For a fixed p, we see that every prime r such that $r \equiv -1 \pmod p$, $r \mid n$, contributes a factor p to $\sigma^*(n)$. Since the number of integers $n < x$ for which $r \mid n$ is $\left[\frac{x}{r}\right]$, it follows that for a given p the number of times the term $\frac{1}{p}$ occurs in the sum S corresponding to each prime $r \equiv -1 \pmod p$ is less than $\left[\frac{x}{r}\right]$. Also, on using the Brun-Titchmarsh estimate for primes in arithmetic progression [6, p. 320] we have

$$\sum_{r \equiv -1 \pmod p} \left[\frac{x}{r}\right] < \frac{c\, x \log\log x}{p}.$$

Hence

$$S < c\, x \log\log x \sum_{p > (\log\log x)^{1+\epsilon}} \frac{1}{p^2} = o(x).$$

Proof of Lemma 3. Given a $p < t$, we see, on using the sieve of Eratosthenes and the fact that

$$\sum_{r \equiv -1 \pmod p} \frac{1}{r} = \infty,$$

that the number of integers $n \le x$ such that n is divisible by at most j primes q of the form $q \equiv -1 \pmod p$, each of them occurring to the first power in n, is $o(x)$, j being an arbitrary positive integer. Hence the number of such integers $n \le x$ is $o(x)$. Since for each such n we have $p^j \mid \sigma^*(n)$, the lemma follows at once.

3. Proof of the theorem. Let η be chosen arbitrarily small and then keep it fixed. We shall then choose t and $\alpha = \alpha(t)$ sufficiently large so that

$$(3.1) \qquad \prod_{p < t} \left(1 + \frac{1}{p^{\alpha}}\right) < 1 + \eta$$

and

$$(3.2) \qquad \prod_{p \ge t} \left(1 + \frac{1}{p^{2}}\right) < 1 + \eta.$$

The latter inequality is possible because of the convergence of $\prod\left(1 + \frac{1}{p^2}\right)$.

Since almost all $n < x$ satisfy Lemmas 1, 2, 3, we have for almost all n,

$$(3.3) \qquad \frac{\sigma_2^*(n)}{\sigma_1^*(n)} \le \prod_{p \le t} \left(1 + \frac{1}{p^{\alpha}}\right) \prod_{p > t} \left(1 + \frac{1}{p^{2}}\right) \cdot \prod_{(\log\log x)^{1-\epsilon} < p < (\log\log x)^{1+\epsilon}} \left(1 + \frac{1}{p}\right),$$

on noting that

$$(3.4) \qquad \sum_{(\log\log x)^{1-\epsilon} < p < (\log\log x)^{1+\epsilon}} \frac{1}{p} < \eta$$

for a suitably chosen $\epsilon = \epsilon(\eta)$.

Combining Lemma 2 and the result (3.4), we get

$$\prod_{\substack{p > t \\ p \mid \sigma^*(n) \\ p^2 \nmid \sigma^*(n)}} \left(1 + \frac{1}{p}\right) < 1 + \eta.$$

It then follows from (3.3) that for almost all n, i.e., except for values of n with density zero,

$$\frac{\sigma_2^*(n)}{\sigma_1^*(n)} < 1 + \eta,$$

and the proof of the theorem is complete. Our theorem implies that $\sigma_2^*(n)/n$ has the same distribution function as $\sigma_1^*(n)/n$.

4. <u>Some remarks and problems</u>. Let $\varphi^*(n)$ be the unitary analogue of Euler's totient function (see E. Cohen [1]). Then $\varphi^*(n)$ has the evaluation

$$\varphi^*(n) = \prod_{p^a \| n} (p^a - 1).$$

Following the method of proof of Theorem 1, we can show that

$$\frac{\varphi_2^*(n)}{\varphi_1^*(n)} \to 1 \qquad (\varphi_1^*(n) = \varphi^*(n))$$

except for a sequence of values of n of density zero. We shall not give the details of proof.

Let $R = R(n)$ be the smallest integer such that $\varphi_R(n) = 1$. This function was first considered by S. S. Pillai [5] who proved that

$$\frac{\log(n/2)}{\log 3} + 1 \le R(n) \le \frac{\log n}{\log 2} + 1.$$

Others who considered this function include Niven [4], Shapiro [7] and Subbarao [8].

Let

$$T(n) = \varphi_1(n) + \varphi_2(n) + \cdots + \varphi_R(n).$$

Since $\varphi_2(n) = o\left(\varphi_1(n)\right)$ for almost all n, and $\varphi_j(n)$ is even for $j \ge 1$, we easily obtain that for almost all n

$$T(n) = (1 + o(1))\varphi(n),$$

so that $T(n) < n$ for almost all n.

There are many problems left about $T(n)$ and we state a few of them below.

Denote by $F(x,c)$ the number of integers $n \le x$ for which $T(n) > cn$. For every $1 < c < 3/2$ we have for every $t > 0$ and $\epsilon > 0$, if $x > x_o = x_o(c,t,\epsilon)$,

$$\text{(4.1)} \qquad \frac{x}{\log x}(\log\log x)^t < F(x,1+c) < \frac{x}{(\log x)^{1-\epsilon}}.$$

This follows easily from Theorem 1 of [2]. Further we have

$$\text{(4.2)} \qquad F(x,1) = \left(c + o(1)\right)\frac{x}{\log\log\log\log x}.$$

The proof of (4.2) can be obtained by the methods used in this paper and by those of [2].

It seems likely that for $1 < c_1 < c_2 < \frac{3}{2}$,

$$\lim_{x\to\infty} F(x,1+c_1)/F(x,1+c_2) = \infty.$$

Put

$$L = \overline{\lim}\, \frac{T(n)}{n}.$$

Trivially $L \le 2$ ($L = 2$ if there are infinitely many Fermat primes). It is easy to show that

$$\overline{\lim}\, \frac{T(2n)}{2n} = 1.$$

We can show that $T(n) > \frac{3n}{2}$ for infinitely many n, which implies $L \ge \frac{3}{2}$. We cannot show that $L > \frac{3}{2}$.

Equation (3) of Theorem 1 of [2] implies that for $c > \frac{3}{2}$ and every $\epsilon > 0$,

$$F(x,c) = o\left(\frac{x}{(\log x)^{2-\epsilon}}\right).$$

Probably,

$$F(x,\tfrac{3}{2}) = o\left(\frac{x}{\log x}\right).$$

but we have not worked out the details.

Some other questions that are still unanswered are the following;

(i) Does $\frac{R(n)}{\log n}$ have a distribution function?

(ii) Does $\frac{R(n)}{\log n}$ approach a limit for almost all n? If this limit exists, is it equal to $\frac{1}{\log 2}$ or $\frac{1}{\log 3}$?

Similar questions arise in the case of the function $R^* = R^*(n)$ defined as the smallest integer such that $\varphi_{R^*}(n) = 1$. Here $\varphi^*(n)$ is the unitary analogue of the Euler totient, introduced by Eckford Cohen [1], which is defined as the multiplicative function for which $\varphi^*(p^k) = p^k - 1$ for all primes p and all positive integers k. We do not even know of any nontrivial estimate for $R^*(n)$. Probably $R^*(n) = o(n^{\epsilon})$ for every $\epsilon > 0$. It is not clear to us at present if $R^*(n) < c \log n$ has infinitely many solutions for some $c > 0$.

REFERENCES

1. E. Cohen, Arithmetic functions associated with the unitary divisors of an integer, Math. Z. 74 (1960), 66-80.

2. P. Erdős, Some remarks on the iterates of the φ and σ functions, Colloq. Math. 17 (1967), 195-202.

3. A. Makowski and A. Schinzel, On the functions $\varphi(n)$ and $\sigma(n)$, Colloq. Math. 13 (1964), 95-99.

4. I. Niven, The iteration of arithmetic functions, Canad. J. Math. 2 (1950) 406-408.

5. S. S. Pillai, On a function connected with $\varphi(n)$, Bull. Amer. Math. Soc. (1929), 837-841.

6. K. Prachar, Primzahlverteilung, Springer-Verlag, Berlin, 1957.

7. H. N. Shapiro, On the iterates of a certain class of arithmetic functions, Comm. Pure Appl. Math. 3 (1950), 259-272.

8. M. V. Subbarao, On a function connected with $\varphi(n)$, J. Madras Univ. B. 27 (1957), 327-333.

DISTRIBUTION OF ADDITIVE AND MULTIPLICATIVE FUNCTIONS

Janos Galambos, Temple University

1. Introduction. An arithmetical function $f(n)$ is called additive if for all coprime m and n,

$$f(mn) = f(m) + f(n), \tag{1}$$

and strongly additive if (1) holds and if for all integers $a \geq 1$ and for any prime q,

$$f(q^a) = f(q). \tag{2}$$

$f(n)$ is called strongly multiplicative if, in addition to (2), the relation

$$f(mn) = f(m)f(n) \tag{3}$$

holds for $(m,n) = 1$.

Let $u(n)$ be an arithmetical function and put

$$F_N(x) = \sum_{\substack{n \leq N \\ u(n) < x}} 1. \tag{4}$$

We say that $u(n)$ has an asymptotic distribution (or limit law), if there is a distribution function $F(x)$ with $F_N(x) \to F(x)$ for all continuity points of the latter. Our aim is to give conditions for an arithmetical function $u(n)$ to have a limit law, with main emphasis on the classes of additive and multiplicative functions. This theory has a wide literature and a good account of works can be found in Kubilius [16] and [18] and Galambos [9] and [11]. The present paper is a continuation of my work in [8] and [9].

The concept of a limit law defined in terms of $F_N(x)$ suggests that our problem is of probabilistic character. As a matter of fact, considering the probability space $S_N = (X_N, \mathcal{C}_N, P_N)$ with $X_N = \{1, 2, \ldots, N\}$ and P_N generating the uniform distribution on the set $\mathcal{C}_N$ of all subsets of X_N, i.e., $P_N(\{i\}) = 1/N$ for all i, any arithmetical function $u(n)$ is a random variable on S_N and

$$P_N(u(n) < x) = F_N(x), \tag{5}$$

hence well known methods of probability theory are applicable to find the limit of $F_N(x)$.

Let $f(n)$ be a real-valued strongly additive function. We have by induction from (1) and (2) that

$$f(n) = \sum_{q|n} f(q) = \sum_{k \geq 1} f(q_k)\, e_k(n) \tag{6}$$

where

$$e_k(n) = \begin{cases} 1 & \text{if } q_k | n, \\ 0 & \text{otherwise.} \end{cases} \tag{7}$$

Here, and in what follows, $q_1 < q_2 < \cdots$ denotes the increasing sequence of prime numbers. By definition

$$P_N(e_k = 1) = (1/N)[N/q_k] = 1/q_k + O(1/N) \tag{8}$$

and similarly

$$P(e_{k_1} = 1, \ldots, e_{k_t} = 1) = 1/q_{k_1} \cdots q_{k_t} + O(1/N). \tag{9}$$

(8) and (9) imply that the random variables $\{e_k\}$ are "almost independent" in the sense of probability theory and (6) thus shows that an additive arithmetical function $f(n)$ is the sum of "almost independent" random variables on S_N. Using Fourier transforms, the continuity theorem of characteristic functions (see Loeve [20], p. 192) yields that there is a limit law for $f(n)$ if the limit, as $N \to +\infty$, of

$$E_N\left[\exp\left(it \sum f(q_k)\, e_k(n)\right)\right] = \frac{1}{N} \sum_{n=1}^{N} e^{itf(n)} \tag{10}$$

exists and if it is continuous at $t = 0$. The almost independence of the e_k's will help in approximating the left hand side of (10) by a product and it will then be easy to obtain conditions on $f(n)$ to guarantee the existence of the limit of the left hand side of (10).

Let $g(n) \neq 0$ be a real-valued strongly multiplicative function. (2) and (3) yield

$$(11) \qquad |g(n)| = \prod_{q|n} |g(q)| = \prod_{k \geq 1} \exp\{e_k(n) \log |g(q_k)|\}$$

and therefore $g(n)$ is the product of "almost independent" random variables. For their investigations the modified Mellin transforms are applied. Here the main tool is the following result of Zolotarev [26]. If the limits, as $N \to +\infty$, of

$$(12) \qquad E_N\left[\exp\left(it \sum_{k \geq 1} e_k(n) \log |g(q_k)|\right)\right] = \frac{1}{N} \sum_{n=1}^{N} |g(n)|^{it}$$

and of

$$(13) \qquad E_N\left[\operatorname{sgn}(g(n))\ |g(n)|^{it}\right] = \frac{1}{N} \sum_{n=1}^{N} |g(n)|^{it} \operatorname{sgn}(g(n))$$

exist and if they are continuous at $t = 0$, then the limit law of $g(n)$ exists and is uniquely determined by the limits of (12) and (13). Note that the left hand sides of (12) and (13) are expressions similar to the left hand side of (10), hence a unified approach is possible to find these three limits. We shall adopt a purely probabilistic approach and find the limits of the left hand sides of (10), (12) and (13) for arbitrary probability spaces, when the random variables corresponding to the e_k's defined in (7) satisfy an "almost independence" requirement. The advantage of this approach is twofold. First of all, limit laws for additive and multiplicative functions are obtained in a single theorem, and secondly, using different probability spaces, we shall have limit laws for arithmetical functions when the argument goes through a given sequence of integers (not necessarily the successive ones). This approach was first applied in Galambos [7] (which I submitted for publication in the summer of 1968) and the more general result to be formulated in the next section is proved in [8]. The application of this general theorem to the distribution problem of multiplicative functions is new and so are some of the theorems on the joint distribution of several arithmetical functions. The result obtained by the method of moments is also formulated here for the first time, though in a vague form it was previously announced by me in [10]. My most recent results, which are to appear in [11] and [12], will be formulated in the last sections. In these, analytical methods are applied.

2. The probabilistic model. Let $S_N = (X_N, G_N, P_N)$ be a sequence of probability spaces and let the random variables $e_{k,N}$, $k = 1, 2, \ldots, s = s(N)$, taking only two values, 0 or 1, be defined on S_N. Put

$$p_{k,N} = P_N(e_{k,N} = 1).$$

We say that the random variables $e_{k,N}$ defined above form a class K if they satisfy the following conditions: there exists a sequence $\{p_k\}$ of real numbers such that $0 \le p_k \le 1$ and

$$p_{k,N} = p_k + O(1/N); \tag{K.1}$$

further, there exists a function $M = M(N) \to +\infty$ with N such that for $1 \le i_1 < \cdots < i_t \le M$,

$$P_N(e_{i_1,N} = 1, \ldots, e_{i_t,N} = 1)$$

$$= p_{i_1} \cdots p_{i_t} + r(M, N; i_1, \ldots, i_t), \tag{K.2}$$

$$|r(M, N; i_1, \ldots, i_t)| \le R(N) \to 0 \quad \text{as} \quad N \to +\infty,$$

where $R(N)$ is independent of the subscripts $i_1, \ldots, i_t$, and the final assumption is that there is a set Z of sequences $\{c_k\}$ of real numbers such that if $\{c_k\} \in Z$, so is every (finite) subset of it and if $\{c_k\} \in Z$, then

$$V_N\Big(\sum_{k=1}^{s} c_k e_{k,N}\Big) \overset{\text{def}}{=} E_N\Big(\Big[\sum_{k=1}^{s} c_k (e_{k,N} - p_k)\Big]^2\Big)$$

$$\le A \sum_{k=1}^{s} c_k^2 p_k + O(1) \quad \text{as} \quad N \to +\infty \tag{K.3}$$

with a suitable constant A not depending on $\{c_k\}$. Here, and in what follows, $E_N(\cdot)$ stands for the expectation of the random variable in the brackets.

Generalizing my result in [7] (the publication of which was delayed by three years), I proved the following theorems in [8].

Theorem 1. *Let* $e_{k,N}$ *be a class* K *and let* Z *be the set occurring in* (K.3). *Let* z_k *be a sequence of complex numbers and put* $x_k = \operatorname{Re} z_k$ *and* $y_k = \operatorname{Im} z_k$. *Suppose that* $\{x_k\}$ *and* $\{y_k\}$ *belong to* Z; *and suppose further that*

(i) $$x_k \le 0,$$

that each of the series

(ii) $$\sum_{|x_k| > 1} p_k, \quad -\sum_{|x_k| \le 1} x_k p_k,$$

(iii) $$\sum_{|y_k| > 1} p_k, \quad \sum_{|y_k| \le 1} y_k p_k, \quad \sum_{|y_k| \le 1} y_k^2 p_k$$

converges, and that

(iv) $$s = o(N).$$

Then

(14) $$\lim_{N=+\infty} E_N\Big[\exp\Big(\sum_{k=1}^{s} z_k e_{k,N}\Big)\Big] = \prod_{k=1}^{+\infty} \Big[1 + (e^{z_k} - 1)p_k\Big].$$

Theorem 2. *Let the conditions of Theorem 1 be satisfied. Let* $u_{1,N}$, $u_{2,N}, \ldots, u_{r,N}$, *be further random variables on* S_N *taking the values* 1 *or* 0, *and suppose that*

(15) $$\lim_{N=+\infty} \sum_{j=1}^{r} |z_{s+j}| \, P_N(u_{j,N} = 1) = 0.$$

Then

$$\lim_{N=+\infty} E_N\Big[\exp\Big(\sum_{k=1}^{s} z_k e_{k,N} + \sum_{j=1}^{r} z_{s+j} u_{j,N}\Big)\Big] = \prod_{k=1}^{+\infty} [1 + (e^{z_k} - 1)p_k].$$

The proofs are simple and short. The major steps in them are as follows. By a sieve type argument we show that (14) holds if on the left hand side we sum only up to an $m \le M$, suitably chosen, and then it is shown that the difference between the limits when we sum up to m or up to s is zero. In this last step, two factors should be emphasized. First of all, by assumption (i), $E_N[\exp(\sum z_k e_{k,N})]$ is bounded by one; on the other hand, the inequality (K.3) applies. These two estimates are to be applied alternatively when the convergence of the series (ii) and (iii) is made use of.

3. Applications to distribution of additive functions. In setting up the probabilistic model of the previous section, we were guided by the observations in the introduction, especially by the relations (6), (8) and (9). Hence the way of applications is now immediate. Let m_j, $j \ge 1$, be an increasing sequence of positive integers and put $X_N = \{m_1, m_2, \ldots, m_N\}$. Let P_N generate the uniform distribution on the set G_N of all subsets of X_N, i.e., $P_N(\{m_j\}) = 1/N$ for all j. Let q_j again be the increasing sequence of prime numbers and define $s = s(N)$ by the relation $q_s \le N < q_{s+1}$. The sequence $e_{k,N}$ is defined as

$$(16) \qquad e_{k,N}(m) = \begin{cases} 1 & \text{if } q_k | m, \\ 0 & \text{otherwise.} \end{cases}$$

Let $M(k_1, k_2, \ldots, k_t; N)$ denote the number of solutions of the congruence

$$m_j \equiv 0 \mod q_{k_1} q_{k_2} \cdots q_{k_t}$$

with $j \le N$ and put $p_{k_1,k_2,\ldots,k_t,N} = M(k_1, k_2, \ldots, k_t; N)/N$. Evidently

$$P_N(e_{k_1,N} = 1, \ldots, e_{k_t,N} = 1) = p_{k_1,\ldots,k_t,N}$$

and

$$f_N(m_j) = \sum_{k \ge 1} y_k e_{k,N}(m_j) t^{-1}$$

is an additive function with $y_k = f_N(q_k) t$ where the argument goes through the elements of X_N. Hence, if m_j is such that the $e_{k,N}$

form a class K, Theorems 1 and 2 are directly applicable, and thus taking $x_k = 0$, they reduce to the limit of the characteristic function of $f_N(m_j) = f(m_j)$. Therefore, the continuity theorem referred to in the introduction, combined with Theorems 1 and 2, yields conditions for the existence of a limit law. This is a very general result, since the question whether the $e_{k,N}$ form a class K for a given sequence m_j has been solved for a very wide class of sequences. As a matter of fact, (K.1) and (K.2) are simple and well known for several sequences and (K.3) is a general form of the Kubilius - Turán inequality (see Kubilius [16] and Galambos [7]). Hence, if $m_j = j$, we have the Erdös theorem (Erdös [6]), if $m_j = Q(j)$, where $Q(x)$ is a positive polynomial with integral coefficients, we have the sufficiency part of the result of Uzdavinys [25]. Taking m_j as the set of integer ideals in a number field, we obtain a theorem similar to that of Erdös for additive functions defined on number fields. The real generality of this section was implicitly shown by Elliott [5] proving that (K.3) is a weaker assumption than the Bombieri form of the large sieve. This fact implies that Theorems 1 and 2 contain the sufficiency parts of Barban [2], Barban, Vinogradov and Levin [3], and Kátai [14] (more exactly, those theorems of these papers which are related to the Erdös theorem). Reformulation of Theorems 1 and 2 for the specific sequences listed above would yield several individual results, interesting in their own.

All those said above can be repeated for the following special additive function. Let $h_1, h_2, \ldots, h_r$ be additive functions and $t_1, t_2, \ldots, t_r$ be real numbers and put

$$f(n) = t_1h_1(n) + t_2h_2(n) + \cdots + t_rh_r(n).$$

If we choose $x_k = 0$ and $y_k = f(q_k)$, the expression on the left hand side of (14) reduces to the multivariate characteristic function of the h's, hence by the continuity theorem of characteristic functions, our theorems imply sufficient conditions for several additive functions to have a joint limit law when the argument n runs through a general sequence of integers. All those special cases which were mentioned at the beginning of this section apply here, hence the result of Kátai [15] is also reobtained (i.e., when n goes through the values of polynomials with integral coefficients).

4. Applications to distribution of multiplicative functions. Let $g(n) \neq 0$ be a strongly multiplicative function. Considering again the probability space described in the first paragraph of the previous section, we assume again that the sequence m_j is such that the random variables $e_{k,N}$ defined in (16) form a class K. With the choice of $x_k = 0$ and $y_k = t \log |g(q_k)|$, the expression on the left hand side of (14) becomes the modified Mellin transform corresponding to (12), and if we choose $x_k = 0$ and

$$y_k = \tfrac{1}{2}\pi(1 - g(q_k)/|g(q_k)|) + t \log |g(q_k)|$$

then the expression in (14) becomes the other Mellin transform corresponding to (13). Simultaneous applications of our theorems to these two cases yield conditions for $g(n)$ to have a limit law. Here, however, we have to distinguish two limit laws. We call a distribution function $F(x)$ symmetric, if for all real x, $F(x) = 1 - F(-x + 0)$. For $g(n)$ to have a symmetric limit law, the transformation (13) should tend to zero, hence Theorems 1 and 2 apply only to the case of nonsymmetric limit laws. Using the notations of the previous section, we have the following

Theorem 3. Let m_j be a sequence of positive integers for which the functions defined in (16) form a class K. Assume that each of the series

$$\sum_{g(q_k) \le \frac{1}{c}} p_k , \qquad \sum_{g(q_k) \ge c} p_k ,$$

$$\sum p_k \log g(q_k) , \quad \sum p_k \log^2 g(q_k)$$

converges for a suitable $c > 1$, where the last two summations are for all k for which $1/c < g(q_k) < c$. Then there is a nonsymmetric distribution function $F(x)$ such that at each of its continuity points

$$\lim_{N=+\infty} \frac{1}{N} \sum_{\substack{k \le N \\ g(m_k) < x}} 1 = F(x).$$

The limit relation above holds also at $x = 0$ and $x = +0$.

Theorem 3 easily follows from Theorems 1 and 2. For $m_j = j$, this result is due to Bakstys [1], in which case the conditions above are also necessary. As a special case, let us mention the choice of $m_j = Q(j)$, where $Q(x) > 0$ is a polynomial with integral coefficients. It is well known that now $p_k = b(q_k)/q_k$, where $b(q)$ is the number of residue classes $\bmod q$ for which $Q(x) \equiv 0 \bmod q$. (K.3) is due to Uzdavinys [24] under the condition that $\lim g(q) = 1$ as $q \to +\infty$. Professor Kubilius tells me that Theorem 3 has been proved for some special sequences by his pupils, but in this general form it appears to be new. The case of symmetric limit law has recently been settled by Galambos [11], which theorem will be returned to in a later section.

5. <u>The class H of Kubilius and the method of moments.</u> Let $f(n)$ be a strongly additive function and put

$$E_N = \sum_{q \le N} \frac{f(q)}{q}, \quad D_N^2 = \sum_{q \le N} \frac{f^2(q)}{q}.$$

We say that $f(n)$ belongs to the <u>class H</u> of Kubilius if $D_N \to +\infty$ and if there is an integer-valued function $r(N) \to +\infty$ and such that $D_{r(N)}/D_N \to 1$ and $\log r(N)/\log N \to 0$. For the class H, Kubilius [16] proved the following general theorem.

<u>Theorem</u> K. <u>In order that for</u> $f(n)$ <u>(belonging to the class</u> H) <u>there be a distribution function</u> $F(x)$ <u>such that for all of its continuity points</u>

$$\lim F_N(E_N + xD_N) = F(x) \quad (N \to +\infty),$$

<u>it is necessary and sufficient that there exist a distribution function</u> $K(u)$ <u>with variance one for which the relation</u>

$$\lim_{N=+\infty} D_N^{-2} \sum \frac{f^2(q)}{q} = K(u)$$

<u>holds for all continuity points of</u> $K(u)$ (<u>summation is for</u> $q \le N$, $f(q) < uD_N$).

Misevicius [21] gave a new proof for Theorem K by the method of moments, i.e., he proved that, putting $f^*(n) = (f(n) - E_N)/D_N$,

$$\lim_{N=+\infty} \frac{1}{N} \sum_{n=1}^{N} [f^*(n)]^k = d_k$$

exists for all k and that d_k is the k-th moment of a distribution function $F(x)$ determined by $K(u)$ (i.e., d_k is the integral of x^k with respect to $F(x)$). By his method, the sufficiency part of Theorem K can be proved for the case when in (4) we replace the sequence of successive integers by a general sequence m_j for which the $e_{k,N}$ of (16) form a class K and if the following conditions are satisfied: (i*) $p_k = w(k)/q_k$ with $w(k)$ bounded, (ii*) replacing $1/q_k$ by p_k in the definitions of E_N and D_N, $f(q) = \mathcal{C}(D_N)$. Very likely, these are not necessary assumptions.

6. Analytical methods. Let $f(n)$ be a strongly additive function and let A_N and B_N be given sequences of real numbers. The problem of finding conditions under which $(f(n) - A_N)/B_N$ has a limit law, i.e., the limit of $F_N(A_N + xB_N)$ exists and is a distribution function, becomes much more difficult if A_N is not assumed to be E_N and/or B_N is not necessarily D_N. The case of $B_N = 1$ was first investigated by Kubilius [17] and necessary and sufficient conditions have recently been obtained by Elliott and Ryavec (announced at a meeting of the American Mathematical Society in Urbana, Illinois in 1970) and by B. V. Levin and also by J. Kubilius (personal communication). The fact that the "best" normalizing constants are not necessarily E_N and D_N has been shown, through several examples, by Levin and Fainleib [19]. The result of Levin and Fainleib [19] does not contain Theorem K, but it is applicable both to functions belonging to the class H and to those outside of it. I recently proved a theorem generalizing Theorem K, which is much more general in two ways. We do not have to assume that $f(n)$ belongs to the class H, neither is any assumption made on the form of A_N and B_N. The method of Halász [13] is used; details are in [12]. In order to formulate our result, we introduce some notations. Put

$$z = 1 + 1/\log N.$$

We say that (R.1) holds with B_N if $B_N \to +\infty$, if there is a real number c such that for all real u with $\frac{1}{2}(\log N)^{-1} \le u-1 \le \frac{3}{2}\log N$, and using the notation $f_c(n) = f(n) - c\log n$,

$$\lim_{N=+\infty} \sum_q q^{-u}\left(1 - \mathrm{Re}\{\exp[it\, f_c(q)/B_N]\}\right) = a(f,t)$$

exists and is independent of u. (R.2) is said to be satisfied if

$$\lim_{N=+\infty} \sum_q (q^{-s} - q^{-z})\ \mathrm{Im}\{\exp[it\, f_c(q)/B_N]\} = 0$$

uniformly on $\{s: 1 + \frac{1}{2}(z-1) \le \mathrm{Re}\, s \le z, \quad |\mathrm{Im}\, s| \le K(z-1)\}$, where K is an arbitrary fixed real number. We now have

<u>Theorem</u> 4. <u>Let</u> $f(n)$ <u>be a real-valued strongly additive function and let</u> B_N <u>be a sequence of positive real numbers satisfying (R.1) and (R.2). Assume that</u>

$$A_N^* = (B_N/t)\left(\sum_q q^{-z}\ \mathrm{Im}\{\exp[it\, f_c(q)/B_N]\} + b(f,t)\right) + o(B_N)$$

<u>is independent of</u> t <u>with a suitable function</u> $b(f,t)$ <u>of</u> t, <u>not depending on</u> N. <u>Then, if both</u> $a(f,t)$ <u>and</u> $b(f,t)$ <u>are continuous at</u> $t = 0$,

$$(f(n) - c\log N - A_N^*)/B_N$$

<u>has a limit law, the characteristic function</u> $w(t)$ <u>of which is given by</u>

$$\log w(t) = a(f,t) + i\, b(f,t)\ .$$

From the method of proof I conjecture that the conditions here are also necessary. The only reason why it is not obvious is that I turned to $f_c(n)$, which is an arbitrary step and which was guided by a recent result of Ryavec [23] concerning Halász' result [13]. If Ryavec's result extends to the generalized form of Halász' result, then the conjecture is right.

I mention here my result on the symmetric limit law for strongly multiplicative functions, which was also obtained by analytic methods. By the result of Zolotarev [26], we need conditions to guarantee the existence of a nonzero limit of (12) which is continuous at $t = 0$, and a zero limit in (13). Since the functions on the right hand sides of (12) and (13) are multiplicative and bounded by one, we can apply the well known result of Delange [4] and Satz 2 of Halász [13], from which necessary and sufficient conditions are obtained in

Galambos [11] under the condition $g(2) \neq -1$. Since this paper is to appear soon, I do not give further details.

For another method to deal with the distribution problem of additive functions see the recent work of Philipp [22].

REFERENCES

1. A. Bakstys, On the asymptotic distribution of multiplicative number theoretic functions (Russian), Litovsk. Mat. Sb. 8 (1968), 5-20.

2. M. B. Barban, Arithmetical functions on "rare" sets (Russian), Trudy Inst. Mat. Akad. Nauk Uzbek. SSR 22 (1961), 1-20.

3. M. B. Barban, A. I. Vinogradov and B. V. Levin, Limit laws for functions of the class H of J. P. Kubilius which are defined on a set of "shifted" primes (Russian), Litovsk. Mat. Sb. 5 (1965), 5-8.

4. H. Delange, Sur les fonctions arithmetiques multiplicatives, Ann. Sci. École Norm. Sup. 78 (1961), 273-304.

5. P. D. T. A. Elliott, The Turán - Kubilius inequality and a limitation theorem for the large sieve, Amer. J. Math. 92 (1970), 293-300.

6. P. Erdős, On the density of some sequences of numbers III, J. London Math. Soc. 13 (1938), 119-127.

7. J. Galambos, Limit distribution of sums of (dependent) random variables with applications to arithmetical functions, Z. Wahrscheinlichkeitstheorie und Verw. Gebiete 18 (1971), 261-270.

8. J. Galambos, A probabilistic approach to mean values of multiplicative functions, J. London Math. Soc. (2) 2 (1970), 405-419.

9. J. Galambos, Distribution of arithmetical functions. A survey, Ann. Inst. H. Poincaré, Sect. B 6 (1970), 281-305.

10. J. Galambos, Probabilistic methods in the theory of numbers, Volume of Contributed Papers, International Congress of Mathematicians, Nice, 1970, p. 154.

11. J. Galambos, On the distribution of strongly multiplicative functions, Bull. London Math. Soc. (to appear).

12. J. Galambos, Integral limit laws for additive functions, (under reference).

13. G. Halász, Uber die Mittelwerte multiplikativer zahlentheoretischer Funktionen, Acta Math. Acad. Sci. Hungar. 19 (1968), 365-402.

14. I. Kátai, On the distribution of arithmetical functions on the set of prime plus one, Compositio Math. 19 (1968), 278-289.

15. I. Kátai, On the distribution of arithmetical functions, Acta Math. Acad. Sci. Hungar. 20 (1969), 69-87.

16. J. Kubilius, Probabilistic methods in the theory of numbers, Amer. Math. Soc. Translations, Vol. 11, Providence, 1964.

17. J. Kubilius, On asymptotic laws of distribution of additive arithmetical functions (Russian), Litovsk. Mat. Sb. 5 (1965), 261-273.

18. J. Kubilius, On the distribution of number theoretic functions, Seminaire Delange-Pisot-Poitou No. 23, 11 (1970), 1-11.

19. B. V. Levin and A. S. Fainleib, Application of some integral equations to questions of number theory (Russian), Uspehi Math. Nauk 22 (135) (1967), 119-197.

20. M. Loeve, Probability Theory, Third Edition, Van Nostrand, Princeton, 1963.

21. G. Misevicius, The use of the method of moments in probabilistic number theory (Russian), Litovsk. Mat. Sb. 5 (1965), 275-289.

22. W. Philipp, Mixing sequences of random variables and probabilistic number theory, Mem. Amer. Math. Soc. (to appear).

23. C. Ryavec, A characterization of finitely distributed additive functions, J. Number Theory 2 (1970), 393-403.

24. R. Uzdavinys, Arithmetical functions on the set of values of integer valued polynomials (Russian), Litovsk. Mat. Sb. 2 (1962), 253-280.

25. R. Uzdavinys, An analogue theorem to that of Erdős and Wintner on the sequences of the values of integer valued polynomials, Litovsk. Mat. Sb. 7 (1967), 329-338.

26. V. M. Zolotarev, General theory of the multiplication of random variables (Russian), Dokl. Akad. Nauk SSSR 142 (1962), 788-791.

OSCILLATION THEOREMS*

Emil Grosswald, Temple University

1. Introduction and notations.

Let

$$F_o(s) = \sum_{n=1}^{\infty} a_n n^{-s}$$

be a Dirichlet series with real coefficients a_n, convergent for some $\sigma \geq \sigma_o \neq \infty$, and set

$$f(x) = \sum_{n \leq x}{}' a_n.$$

Here and in what follows the dash signifies that for integral x the term a_x is counted half, so that $f(x) = \frac{1}{2}\{f(x+0) + f(x-0)\}$ holds for all real x. Let $g(x)$ be a real-valued function, which satisfies the asymptotic equality $f(x) \sim g(x)$ for $x \to \infty$, and denote by $u(x)$ a positive monotonic function. By an oscillation theorem we shall understand a statement to the effect that each of the inequalities

$$(1) \qquad f(x) - g(x) > Cu(x), \quad f(x) - g(x) < -Cu(x)$$

holds for an appropriate constant $C \geq 0$ and for infinitely many values $x = x_n$ with $x_n \to \infty$. Similar sets $\{x_n\}$ will occur often and, to make compact statements easier, we define X-sets as follows:

Definition 1. A denumerably infinite set $\{x_n\}$ of real numbers with $x_{n+1} > x_n$ and $\lim_{n\to\infty} x_n = +\infty$ is called an X-set.

There will be no ambiguity if instead of (1) we write simply $f(x) - g(x) \gtrless \pm Cu(x)$, in order to indicate that both inequalities (1) hold, each on some (naturally different) set and the oscillation theorems will be stated usually in the form:

"Each of the two inequalities $f(x) - g(x) \gtrless \pm Cu(x)$ holds on some X-set."

---

*This research was partially supported by NSF Grant GP-23170.

If $C = 0$, this simply means that the difference $f(x) - g(x)$ changes signs infinitely often. In many of these theorems it can be shown that the corresponding X-sets can be chosen so that they consist only of positive integers, but this fact will neither be proven, nor be used.

The functions whose oscillations will be considered are $\pi(x) - \mathrm{li}\, x$, $\pi(x;k,a) - \pi(x;k',a')$ for $\varphi(k) = \varphi(k')$, $M(x)$, $L(x) - x^{\frac{1}{2}}(\zeta(\frac{1}{2}))^{-1}$, and $N(x) - x^{-\frac{1}{2}}(\zeta(\frac{1}{2}))^{-1}$.

The symbols used here have their usual meaning. Z will be used for the integers, Z^+ for the positive integers, ϕ for the empty set,

$$\pi(x) = \sum_{p \le x}{}' 1, \quad \mathrm{li}\, x = \int_2^x \frac{dy}{\log y},$$

and

$$\pi(x;k,a) = \sum_{\substack{p \le x \\ p \equiv a \pmod k}}{}' 1.$$

We shall also need

$$f_o(x) = \sum_{m \ge 1} m^{-1}\pi(x^{1/m})$$

and

$$f_o(x;k,a) = \sum_{m \ge 1} m^{-1}\pi(x^{1/m};k,a^{1/m}), \text{ with } \pi(y;k,a^{1/m}) = \sum_{\substack{p \le y \\ p^m \equiv a \pmod k}}{}' 1.$$

Next, Euler's function

$$\varphi(k) = \sum_{\substack{m=1 \\ (m,k)=1}}^{k} 1;$$

$$M(x) = \sum_{n \le x}{}' \mu(n),$$

with $\mu(n) = 0$ if n is not squarefree, $\mu(n) = (-1)^r$ for $n = p_1 p_2 \cdots p_r$ $(p_i \neq p_j$ if $i \neq j)$;

$$L(x) = \sum_{n \leq x}{}' \lambda(n)$$

with $\lambda(n) = (-1)^r$ for $n = p_1^{a_1} p_2^{a_2} \cdots p_m^{a_m}$, $\sum_{j=1}^{m} a_j = r$; and

$$N(x) = \sum_{n \leq x}{}' \lambda(n)/n.$$

Riemann's zeta function is defined by

$$\zeta(s) = \sum_{n=1}^{\infty} n^{-s} \qquad (s = \sigma + it;\ \sigma, t \text{ real})$$

for $\sigma > 1$, by analytic continuation otherwise.

Most of the results to be stated are known, or are easily obtained, in case the Riemann Hypotheses (RH) fails to hold. Therefore, except for a specific statement to the contrary, in what follows we shall assume without further mention the validity of the RH, i.e., we assume that $\zeta(s) = 0$ holds only for $s = -2m$ $(m \in Z^+)$, or for $s = \rho_n = \frac{1}{2} + i\gamma_n$. The set of the γ_n's will be denoted by Γ and the multiplicity of the zero ρ by m_ρ.

To any given modulus $k \in Z^+$, there are $\varphi(k)$ distinct characters $\chi_j(n)$ $(j = 1, 2, \ldots, \varphi(k))$. These will be numbered following the method of R. Spira ([14]; see also [3]). If the group of prime residues modulo k is cyclic, this method reduces to the following: Let g be the smallest primitive root modulo k, and let $n \equiv g^r \pmod{k}$; then $\chi_j(n) = e^{2\pi ijr/\varphi(k)}$. This is the case, in particular, when $k = p$, a prime; then $\varphi(p) = p-1$ and there is, besides the principal character $\chi_0(n)$, exactly one other real character, corresponding to $j = (p-1)/2$ and $\chi_{(p-1)/2}(n) = \left(\frac{n}{p}\right)$, the Legendre symbol of quadratic residuacy. Also in the more general case, the Jacobi symbol $\left(\frac{n}{k}\right)$ is a real character, but not necessarily the only one. In case $k = p$, the remaining $p-3$ characters are complex and split into $(p-3)/2$ couples of complex conjugate ones, with $\bar{\chi}_j(n) = \chi_{p-j-1}(n)$. In case $p \equiv 1 \pmod 4$, the two characters with $j = (p-1)/4$ and its conjugate have the property that $\chi_j(n) = \pm i$ if

$\left(\frac{n}{p}\right) = -1$, and $\chi_j(n) = \pm 1$ if $\left(\frac{n}{p}\right) = +1$. To each character corresponds an L-function, defined for $\sigma > 1$ by

$$L(s,\chi_j) = L(s;k,\chi_j) = \sum_{n=1}^{\infty} \chi_j(n) n^{-s}.$$

The complex zeros of $L(s,\chi_j)$, ordered by increasing ordinates, will be denoted by $\rho_n(\chi_j)$. We shall assume (except for an express statement to the contrary, as in Theorem 5 and its proof) that $\rho_n(\chi_j) = \frac{1}{2} + i\gamma_n(\chi_j)$.

In general, all mention of parameters will be kept at a minimum; χ_j, k, etc., will be suppressed, unless needed to avoid ambiguities. The multiplicity of any ρ as a zero of $L(s,\chi_j)$ will be denoted by $m_\rho(\chi_j)$. The possibility that $L(\rho,\chi_q) = L(\rho,\chi_j) = 0$ for $j \neq q$ is not ruled out. For L-series $L(s;k,\chi)$ and $L(s;k',\chi')$, with $\varphi(k) = \varphi(k')$ and integers a, a' with $(a,k) = (a',k') = 1$, the <u>composite multiplicities</u>

$$m_\rho = m_\rho(a,a') = \sum_{\chi} \bar{\chi}_j(a) m_\rho(\chi_j) - \sum_{\chi'} \bar{\chi}'_j(a') m_\rho(\chi'_j)$$

will occur. The functions $L(s,\chi_j)$ may have also real zeros $\sigma_r(\chi_j)$ of multiplicities $m_{\sigma_r}(\chi_j)$ and we consider also the corresponding composite multiplicities

$$m_\sigma = m_\sigma(a,a') = \sum_{j} \bar{\chi}_j(a) m_\sigma(\chi_j) - \sum_{j'} \bar{\chi}'_{j'}(a') m_\sigma(\chi'_{j'}).$$

The principle of the method used and some of the basic theorems are presented in Section 2. The main results are stated in Section 3. Section 4 contains some auxilliary results. The following Sections 5-8 contain the proofs of the theorems. Section 9 contains certain numerical results, and a discussion of the computational procedure.

I wish to express my thanks to Professors H. Diamond and H. Stark, who showed me the manuscripts of [2] and [15] respectively, before these were published, and to Professors P. Hagis and R. Spira who performed the numerical work discussed in Sections 8 and 9.

2. The principle of the method, some theorems and some conjectures. The fundamental idea of the proofs goes back to Landau [11]. By Perron's theorem,

$$f(x) = \frac{1}{2\pi i}\int_{(c)} F_0(s)x^s s^{-1}ds,$$

where the integral is taken along a vertical line of abscissa c, appropriately chosen to insure the convergence of the integral. By inversion one obtains the Mellin integral

$$s^{-1}F_0(s) = \int_0^\infty f(x)x^{-s-1}dx.$$

If we set

$$\int_0^\infty g(x)x^{-s-1}dx = G(s) \quad \text{and} \quad \int_0^\infty u(x)x^{-s-1}dx = U(s),$$

we obtain

$$(2) \quad \int_0^\infty \{f(x) - g(x) - Cu(x)\}x^{-s-1}dx = s^{-1}F_0(s) - G(s) - CU(s) = F(s),$$

say. If for $|C| < C_0$ (and regardless of the sign of C), $F(s)$ can be continued as a holomorphic function up to $\sigma > \theta$, but not into any half-plane $\sigma > \theta - \epsilon$ for $\epsilon > 0$, and if $F(s)$ is holomorphic at $s = \theta$ then Landau's theorem [11] yields (1).

The success of the investigations of [4] depended upon the following extension of Landau's theorem.

Theorem A. (See [5]). Let $f(x)$ be real for real $x \geq x_0$ and suppose that $\int_{x_0}^\infty f(x)x^{-s}dx$ converges for $\sigma \geq \sigma_0$, and represents there a function $F(s)$, holomorphic for $\sigma > \theta$, but not for $\sigma > \theta - \epsilon$, if $\epsilon > 0$. Suppose also that $\limsup_{\sigma \to \theta+} F(\sigma) = +\infty$ and that there exists a $t \neq 0$, such that, for $\sigma \to \theta+$, $|F(\sigma+it)|$ is "bigger" than $|F(\sigma)|$, in the sense that either $\limsup_{\sigma \to \theta+} |F(\sigma+it)/F(\sigma)| > 1$, or $\limsup_{\sigma \to \theta+} \{|F(\sigma+it)| - |F(\sigma)|\} = +\infty$ holds. Then there exists an X-set on which $f(x)$ changes signs.

For the application of Theorem A, one observes that $s^{-1}F(s)$ has singularities on $\sigma = \theta$. One then chooses $u(x)$ and C in such a way, that $\limsup_{s \to (\theta+it)+} |s^{-1}F_o(s) - G(s)|$ should exceed $\limsup_{s \to \theta+} |CU(s)|$ in the sense of Theorem A. So, e.g., if $s^{-1}F(s) - G(s)$ has poles of first order at $\theta + it_j$ $(j = 1, 2, \ldots)$, with respective residues a_j, one will choose $u(x)$ so that $U(s)$ should have a pole of first order at $s = \theta$, with residue equal to one and then take

$$|C| < \sup_j |a_j|. \tag{3}$$

The purpose of this paper is to improve and extend some of the results of [4]. This has been rendered possible, by a corresponding sharpening of Theorem A, due to H. Diamond [2] and stated here as Theorem B. It is convenient to introduce first the concepts of N-independence and independence in (or relative to) a set.

Definition 2. Let $T = \{t_i \mid i \in Z^+\}$ be a (finite or infinite) set of (real or complex) numbers and let $S \subset T$ be a subset of T, consisting of J elements. We shall say that S is N-independent in T if $\sum_{j=1}^{J} n_j s_j = t \in T$, with $n_j \in Z$, $|n_j| \le N$ and $s_j \in S$ is possible only trivially, i.e., with $|n_j| = 1$ for exactly one $j = j_o$ and $n_j = 0$ for $j \ne j_o$.

Theorem B. (H. Diamond, see [2]). Let $f(x)$ be a real-valued function, absolutely integrable on compact subsets of $[0,\infty)$, set $F(s) = \int_0^\infty f(x)e^{-sx}dx$ and assume that the integral converges for $\sigma > 0$. Assume also that $F(s)$ can be continued as a meromorphic function to an open set of the complex plane, containing the imaginary axis. Let $T = \{t_j \mid j \in Z^+\}$ be a non-empty set (finite or denumerable) of real numbers and assume that $F(s)$ has only the poles $s_j = it_j$ on the imaginary axis, that all are simple and with residues a_j (a_o is the residue at $s = 0$; $a_o = 0$ is not excluded). Suppose that there exists a subset $S \subset T$, N-independent in T and such that

(4) $$2 \sum_{s_j \in S} |a_j| > N^{-1}(N+1)|a_o|.$$

Then there exists no real number y, *such that* $f(x)$ *(or any* L^1*-equivalent function) is of a constant sign on* $[y,\infty)$.

We shall need also the following slight extension of Theorem B:

Theorem C. *Let* $f(x)$ *and* $F(s)$ *be defined as in Theorem B, and assume that* $F(s)$ *has only the singularities* $s_j = it_j$ *on the imaginary axis. Let the left complex half-plane be cut along horizontal lines starting at* $s = it_j$ *and assume that, for some* $\epsilon > 0$, $F(s)$ *can be continued as a meromorphic function to* $\sigma > -\epsilon$ *in the cut plane. We also assume that there exist constants* a_j *such that* $F(\sigma+it_j) - a_j \log \sigma$ *stays bounded in a neighborhood of* $s = it_j$ *for all* j. *Then the conclusion of Theorem B still holds. Specifically, if* $S \subset T$ *is N-independent in* T *and (4) holds, then there exists no real number* y, *such that* $f(x)$ *should be of constant sign on* $[y,\infty)$.

A proof of Theorem C is sketched in Section 4.

The use of Theorems B or C, instead of Theorem A, permits us to replace (3) by

(5) $$|c| < 2N(N+1)^{-1} \sum_{i \in J} |a_j|,$$

where j ranges over a subset of the integers, such that the corresponding set $\{t_j\}$ of ordinates is N-independent in the set $T = \{t_j | j \in Z^+\}$. In particular, if there exists a set $S = \{t_{j_n}\}$, $S \subset T$, and independent in T, then one may let $N \to \infty$ and (5) becomes

$$|c| < 2 \sum_{t_j \in S} |a_j|.$$

In the sequel we shall refer also to the following two conjectures, concerning the set $\Gamma = \{\gamma\}$ of imaginary parts of the zeros of $\zeta(s)$.

Conjecture C. $\Gamma = \Gamma_1 \cup \Gamma_2$, *where* Γ_1 *is finite and* Γ_2 *is independent in* Γ.

Conjecture C_o. $\Gamma = \Gamma_1 \cup \Gamma_2$ *where* Γ_1 *is finite and every finite subset of* Γ_2 *is* 1-*independent in* Γ.

Let $\tilde{\Gamma} = \bigcup_{\chi} \Gamma(\chi)$, where $\Gamma(\chi) = \{\gamma \mid L(\beta+i\gamma,\chi) = 0,\ \gamma \neq 0\}$ and the union is taken over a finite set of characters, not necessarily to the same modulus. One can then formulate an extension of Conjecture C_o as follows:

Conjecture $\tilde{C}_o$. $\tilde{\Gamma} = \tilde{\Gamma}_1 \cup \tilde{\Gamma}_2$, *where* $\tilde{\Gamma}_1$ *is finite and every finite subset of* $\tilde{\Gamma}_2$ *is* 1-*independent in* $\tilde{\Gamma}$.

3. *Main results*.

Theorem 1. (i) *Conjecture* C_o *implies that*

$$\lim_{x\to\infty} {}^{\sup}_{\inf} (\pi(x) - \mathrm{li}\, x)x^{-\frac{1}{2}} \log x = \pm\infty.$$

(ii) *Without any conjecture, but upon verification that the ordinates of the first 30 nontrivial zeros of* $\zeta(s)$ *are* 5-*independent in* Γ, $\pi(x) - \mathrm{li}\, x$ *changes signs on an* X-*set*.

Remarks. 1. The result stated under (i) and which assumes C_o is weaker than the known, unconditional result of Littlewood [6], [12] that there exists a $c_o > 0$, such that

$$\lim_{x\to\infty} {}^{\sup}_{\inf} (\pi(x) - \mathrm{li}\, x)x^{-\frac{1}{2}} \log x(\log\log\log x)^{-1} \gtrless \pm c_o.$$

2. The verification of the 5-independence of $\Gamma_1 = \{\gamma_n \mid 1 \le n \le 30\}$ involves the checking of $((2N+1)^n-1)/2 = \frac{1}{2}(11^{30}-1)$ sums.

Theorem 2. (i) *If any of the zeros* $\rho = \frac{1}{2} + i\gamma$ *of* $\zeta(s)$ *is multiple, then*

$$\lim_{x\to\infty} {}^{\sup}_{\inf} M(x)(x^{\frac{1}{2}} \log x)^{-1} \gtrless \pm c$$

for some $c > 0$.

(ii) <u>If all zeros</u> $\rho = \frac{1}{2} + i\gamma$ <u>of</u> $\zeta(s)$ <u>are simple and Conjecture</u> C_0 <u>is valid, then</u>

$$\lim_{x\to\infty} \begin{matrix}\sup\\ \inf\end{matrix} M(x)x^{-\frac{1}{2}} = \pm\infty .$$

(iii) <u>Unconditionally, Mertens' Conjecture that</u> $|M(x)| < \sqrt{x}$ <u>is disproven, pending verification of the 13-independence of the imaginary parts of the first 75 zeros</u> $\rho_n = \frac{1}{2} + i\gamma_n$ <u>of</u> $\zeta(s)$.

(iv) <u>The verification of the 1-independence of</u> $\Gamma_1 = \{\gamma_n | 1 \le n \le 20\}$ <u>in</u> Γ <u>is sufficient to disprove</u>

$$|M(x)| < \frac{1}{3}\sqrt{x} .$$

<u>Remarks</u>. 1. It seems unlikely that $\zeta(s)$ should have multiple zeros.

2. The verification of the 13-independence of the first 75 zeros of $\zeta(s)$ involves the checking of $(27^{75}-1)/2$ sums.

3. The verification of the 1-independence of the first 20 zeros has been done at least partially in [1].

4. Mertens' Conjecture has been disproven by Saffari [13] modulo more extensive computations. The mathematical folklore credits also W. Jurkat with a similar result (no publications).

5. Ingham [9] proved (ii) under the stronger Conjecture C. In the present form, (ii) has recently been proven by Bateman et al. in [1].

<u>Theorem</u> 3. (i) <u>Conjecture</u> C_0 <u>implies that</u>

$$\lim_{x\to\infty} \begin{matrix}\sup\\ \inf\end{matrix} \left\{L(x) - x^{\frac{1}{2}}(\zeta(\tfrac{1}{2}))^{-1}\right\} x^{-\frac{1}{2}} = \pm\infty .$$

(ii) <u>Without any unproven conjecture, the 16-independence in</u> Γ <u>of the set of imaginary parts of the first 13 zeros of</u> $\zeta(s)$ <u>implies that</u> $L(x)$ <u>changes signs on an X-set</u>.

<u>Remarks</u>. 1. The first statement has already been obtained by Bateman et al. [1]. With the stronger Conjecture C it is due to Ingham [9].

2. The verification of the 16-independence of the first 13 imaginary parts of zeros of $\zeta(s)$ in Γ requires the checking of

$(33^{13}-1)/2$ sums. In fact, Haselgrove showed in [7] that $L(x)$ changes signs on an X-set.

Theorem 4. (i) *Conjecture* C_0 *implies that*

$$\lim_{x\to\infty} {\sup \atop \inf} \left\{N(x) + x^{-\frac{1}{2}}(\zeta(\tfrac{1}{2}))^{-1}\right\} x^{\frac{1}{2}} = \pm\infty.$$

(ii) *The* 16-*independence in* Γ *of the first 13 imaginary parts of zeros of* $\zeta(s)$ *implies that* $N(x)$ *changes signs on an* X-*set*.

Remark. Haselgrove has shown in [7] that $N(x)$ changes signs on an X-set.

In the following Theorem 5 we do not necessarily assume the extended Riemann Hypothesis. We set

$$\theta = \max_{\chi,\chi'} \sup \{\beta | \ L(\beta+it,\chi)L(\beta+it,\chi') = 0, \quad t \neq 0\};$$

here the characters χ and χ' are defined mod k and mod k', respectively, with k not necessarily equal to k', but with $\varphi(k) = \varphi(k')$.

For $(a,k) = (a',k') = 1$, the (possibly empty) set Σ of real zeros σ, $0 < \sigma < 1$, of the $2\varphi(k)$ functions $L(s,\chi)$ and $L(s,\chi')$ has a subset Σ_1, $\Sigma \supset \Sigma_1 = \{\sigma | \ \sigma \in \Sigma, m_\sigma(a,a') \neq 0\}$, where $m_\sigma(a,a')$ is the composite multiplicity of σ as defined in Section 1. We now set $\sigma_M = \max_{\sigma \in \Sigma_1} \sigma$. We also abbreviate the expression $\pi(x;k,a) - \pi(x;k',a')$ by $D(x)$, whenever there is no danger of confusion. We denote the number of solutions (distinct modulo k) of $x^2 \equiv 1 \pmod k$ by $n(k)$ and set $n_0 = \max\{n(k), n(k')\}$.

In Theorem 5, and also later, occur sums of the form $\sum_{\gamma \in S} G(\rho)$. These are understood to be taken over sets of zeros ρ, such that the corresponding imaginary parts γ form a set either independent, or N-independent in Γ, or $\tilde{\Gamma}$.

Theorem 5. *For given integers* k, $k' > 1$ *and* a, a' *such that* $(a,k) = (a',k') = 1$, *the following statements hold:*

(i) *If* $\sigma_M > \theta$, *then* $D(x)$ *is ultimately of a constant sign.*

(ii) *If* $\sigma_M < \theta$, $\theta > \frac{1}{2}$ *and* $\{\rho \mid \rho = \theta + i\gamma, m_\rho \neq 0\} \neq \phi$, *then there exists a constant* $c > 0$, *such that both inequalities*

$$D(x) \gtrless \pm cx^\theta / \log x$$

hold on X-*sets.*

(iii) *If* $\sigma_M < \theta = \frac{1}{2}$ (*by the functional equation of* $L(s,\chi)$ *this actually implies* $\Sigma_1 = \phi$) *and* $n(k)\left(\frac{a}{k}\right) = n(k')\left(\frac{a'}{k'}\right)$, *then* $D(x)$ *changes signs on an* X-*set; the condition on* $n(k)$ *and* $n(k')$ *can be omitted, if Conjecture* C_o *holds.*

(iv) *If* $\sigma_M < \theta = \frac{1}{2}$, *denote by* S *any* N-*independent subset of* $\tilde{\Gamma}$. *If such* $S \subset \tilde{\Gamma}$ *exists, so that*

$$\sum_{\gamma \in S} \frac{m_\rho(a,a')}{|\rho|} > (2N)^{-1}(N+1)n_o, \tag{6}$$

then $D(x)$ *changes sign on an* X-*set.*

(v) *In the particular case of (iv), when* $k = k' = p$, *a prime,* $n_o = 2$; *and, if also all zeros* ρ *are simple, (6) holds provided that*

$$\sum_{\gamma \in S} \left|\frac{\bar{\chi}(a) - \bar{\chi}(a')}{\rho}\right| = \sum_{\gamma \in S} \left|\frac{\chi(a) - \chi(a')}{\rho}\right| > 1 + \frac{1}{N}. \tag{7}$$

(vi) $D(x)$ *changes signs on* X-*sets for all primes* $p < 20$ *and all* $(a,p) = (a',p) = 1$, $a \not\equiv a' \pmod p$.

Remarks. 1. The result that $D(x)$ changes signs on X-sets is known for $k = 4$ (Hardy and Littlewood [6]); $k = 3, 6, 8$ (Knapowski and Turán [10]); $k = 43, 67, 163$ [4]. All remaining primes $p < 20$ are discussed in the present paper, Section 8.

2. After completion of the present work, but before that of the final draft, it has come to my attention that H. Stark has settled the case $p = 5$ in [15]. Although the case $a = 4$, $a' = 3$ is formally still left open, it is clear that it can be taken care of by the same method as that used for $a = 4$, $a' = 2$. Consequently, there is no point in completing for $p = 5$ the numerical computations outlined in

Section 8. The corresponding work for $p = 7, 11, 13, 17$ and 19 has been performed by Professor P. Hagis, on the computer CDC 6400 of Temple University. Professor R. Spira kindly performed some needed additional computations at the Michigan State University Computing Center.

4. Auxiliary results.

4.1. Proof of Theorem C. By using properties of the Fejer kernel, H. Diamond [2] proved that if $f(x) \geq 0$ for $y \leq x < \infty$, then, with the notations of Section 1,

$$(8) \qquad F(\sigma) + 2N(N+1)^{-1} \sum_{j=1}^{J} \operatorname{Re}\{F(\sigma+it_j)e^{ic_j}\} + \sum e_n F(\sigma+i\tau_n) \geq 0.$$

Here the last sum is finite, $\tau_n \notin T$ and the c_j's are arbitrary real numbers, that we can dispose of later. In establishing (8), no use was made of any property of $F(s)$ in $\sigma \leq 0$; in particular, the meromorphic character of $F(s)$ played no role. If we recall that $F(s)$ is holomorphic at $s = i\tau_n$, (8) yields

$$a_o \log\sigma + 2N(N+1)^{-1} \sum_{j=1}^{J} \operatorname{Re}\{a_j \log\sigma \cdot e^{ic_j}\} > -B,$$

with some finite, real constant B. We now take $c_j = \pi - \arg a_j$, so that $a_j e^{ic_j} = -|a_j|$ and obtain

$$a_o - 2N(N+1)^{-1} \sum_{j=1}^{J} |a_j| > -B/\log\sigma .$$

We now let σ approach zero, $\sigma \to 0+$ and obtain

$$a_o \geq 2N(N+1)^{-1} \sum_{j=1}^{J} |a_j|,$$

which contradicts (4) and this finishes the proof of the theorem.

4.2. Some integrals. For $\sigma = \operatorname{Re} s > 1$,

(i) $$\int_1^\infty x^{-s} \log x \, dx = (s-1)^{-2}.$$

Proof. Integration by parts.

(ii) $$\int_e^\infty (x^s \log x)^{-1} dx = -\log(s-1) + g(s),$$

with $g(s)$ an entire function.

Proof. Set $y = \log x$; then

$$\int_1^\infty e^{-y(s-1)} \frac{dy}{y} = \int_{s-1}^\infty e^{-u} \frac{du}{u} = \int_{s-1}^\infty \frac{du}{u} + \int_{s-1}^1 \frac{e^{-u}-1}{u} du + \int_1^\infty e^{-u} \frac{du}{u}$$

$$= \log \frac{1}{s-1} + g(s),$$

where $g(s) = C - \sum_{k=1}^\infty \frac{(1-s)^k}{k!k}$ is an entire function

$(C = \int_1^\infty e^{-u} \frac{du}{u} + \sum_{k=1}^\infty \frac{(-1)^k}{k!k})$.

(iii) $$\int_e^\infty \operatorname{li} x \cdot x^{-s-1} dx = \frac{1}{s}\left\{ \log \frac{1}{s-1} + g_1(s) \right\}$$

with $g_1(s)$ an entire function.

Proof. $$\int_e^\infty \left(\left\{ \int_2^e + \int_e^x \right\} \frac{dy}{\log y} \right) x^{-s-1} dx$$

$$= \int_e^\infty c_1 x^{-s-1} dx + \int_e^\infty \left\{ \int_e^x \frac{dy}{\log y} \right\} x^{-s-1} dx.$$

Here,

$$c_1 = \int_2^e \frac{dy}{\log y} \quad \text{and} \quad \int_e^\infty c_1 x^{-s-1} dx = c_1 s^{-1} e^{-s}.$$

Next, using also (ii),

$$\int_e^\infty \left\{ \int_e^x \frac{dy}{\log y} \right\} x^{-s-1} dx$$

$$= \int_e^\infty \frac{1}{\log y} \left\{ \int_y^\infty x^{-s-1} dx \right\} dy$$

$$= \frac{1}{s} \int_e^\infty (y^s \log y)^{-1} dy$$

$$= \frac{1}{s} \left\{ \log \frac{1}{s-1} + g_1(s) \right\}.$$

4.3. The verification of the N-independence of a set consisting of J elements requires the checking of $(2N+1)^J$ sums. Of these one is zero and the others fall into pairs of same absolute value and opposite signs. L-series formed with complex conjugate characters have, however, complex conjugate zeros; it is, therefore, sufficient to consider the $\{(2N+1)^J - 1\}/2$ positive sums and check them against the γ's belonging to all L-series of given modulus.

5. Proof of Theorem 1. In (2) set $f(x) = f_o(x)$, $g(x) = \text{li}\, x$, and $u(x) = x^{\frac{1}{2}}/\log x$. By using 4.2(ii) and (iii) and taking into account that $f_o(x) = \text{li}\, x = 0$ for $x < 2$, we obtain

$$\int_2^\infty \left\{ f_o(x) - \text{li}\, x - cx^{\frac{1}{2}} (\log x)^{-1} \right\} x^{-s-1} dx$$

$$= \frac{1}{s} \left\{ \log (s-1) \zeta(s) + g(s) \right\} - c \log \frac{1}{s-\frac{1}{2}}$$

with $g(s)$ an entire function.

Theorem C applies with $a_o = -c$, $a_j = m_{\rho_j}/\rho_j$, and (5) becomes

(9) $$|c| < 2N(N+1)^{-1} \sum_{\gamma \in S} \frac{m_\rho}{|\rho|} .$$

If Conjecture C_o holds, one may take $N = 1$ and use the divergence of the series $\sum |\rho|^{-1}$ to obtain the result that

$$f_o(x) - \operatorname{li} x - cx^{\frac{1}{2}}(\log x)^{-1}$$

changes signs on an X-set, regardless of the sign, or size of c. In particular, for $c > 0$, arbitrarily large, each of the inequalities

$$f_o(x) - \operatorname{li} x \gtrless \pm cx^{\frac{1}{2}}/\log x$$

holds on an X-set. Finally, if we replace $f_o(x)$ by $\pi(x) + x^{\frac{1}{2}}(\log x)^{-1} + O(x^{\frac{1}{2}}(\log x)^{-2})$, we obtain assertion (i) of the theorem.

A hand computation based on [8] leads to $\sum_{n=1}^{30} |\rho_n|^{-1} > .601$. Using this and $N = 5$ in (9), we obtain $2N(N+1)^{-1} \sum_{n=1}^{30} \frac{1}{|\rho_n|}$ $> \frac{2.5}{6}(.601) > 1.001$ and we may select a $c = 1+v > 1$ $(0 < v < .001)$, satisfying (9). Consequently, both inequalities

$$\pi(x) - \operatorname{li} x \gtrless -\frac{x^{\frac{1}{2}}}{\log x} \pm (1+v)\frac{x^{\frac{1}{2}}}{\log x} + O\left(\frac{x^{\frac{1}{2}}}{\log^2 x}\right)$$

hold on X-sets and this finishes the proof of Theorem 1(ii). Remark 2 is justified by 4.3.

6. <u>Proof of Theorems 2, 3, and 4</u>. In (2) set $f(x) = M(x)$, $g(x) = 0$, $u(x) = x^{\frac{1}{2}}\log x$. Using 4.2(i) and $M(x) = 0$ for $0 < x < 1$, we obtain

$$\int_1^\infty (M(x) - cx^{\frac{1}{2}}\log x)x^{-s-1}dx = \frac{1}{s\zeta(s)} - \frac{c}{(s-\frac{1}{2})^2} = F(s).$$

Let $\rho = \frac{1}{2} + i\gamma$ be a zero of $\zeta(s)$ of multiplicity at least 2 with principal part $\frac{a}{s-\rho} + \frac{b}{(s-\rho)^2} + \cdots$. Then

$$\lim_{\sigma \to \frac{1}{2}+} |F(\sigma+i\gamma)(\sigma-\tfrac{1}{2})^2| \geq |b|, \quad \lim |F(\sigma)(\sigma-\tfrac{1}{2})^2| = c,$$

and for any c with $|c| < |b|$, it follows by Theorem A that both inequalities

$$M(x) \gtrless \pm cx^{\frac{1}{2}}\log x$$

hold on X-sets. Theorem 2(i) is proven.

If all zeros of $\zeta(s)$ are simple, we repeat the reasoning with $u(x) = x^{1/2}$ instead of $x^{1/2}\log x$ and obtain

$$\int_1^\infty (M(x) - cx^{1/2})x^{-s-1}dx = \frac{1}{s\zeta(s)} - \frac{c}{s-\frac{1}{2}} = F(s).$$

By Theorem B with $a_o = -c$ and $a_j = \operatorname{Res}\{(s\zeta(s))^{-1}\}_{s=\rho}$ $= (\rho_j\zeta'(\rho_j))^{-1}$, it follows that $M(x) \gtrless \pm cx^{1/2}$ will hold on X-sets for any c, for which

$$|c| < 2N(N+1)^{-1} \sum_{\gamma \in S} |\rho\zeta'(\rho)|^{-1}.$$

If C_o holds, we select $N = 1$ and recall that (see, e.g., [16], p. 318) $\sum |\rho\zeta'(\rho)|^{-1}$ diverges (clearly, even after a finite set of zeros has been omitted). It follows that we may select $|c|$ arbitrarily large and that proves (ii). We easily check, using [8], that $\sum_{n=1}^{20} |\rho\zeta'(\rho)|^{-1} > .3576$ and $\sum_{n=1}^{75} |\rho\zeta'(\rho)|^{-1} > .5387$, so that

$\frac{2.13}{14} \sum_{n=1}^{75} |\rho\zeta'(\rho)|^{-1} > 1$ and $\frac{2.1}{2} \sum_{n=1}^{20} |\rho\zeta'(\rho)|^{-1} > 1/3$ and this finishes the proof of the theorem (see [13], p. 1097-2000 for the insufficiency of $1 \le n \le 30$).

The proofs of Theorems 3 and 4 are entirely analogous and similar to that of Theorem 2. Equation (2) becomes in these cases

$$\int_1^\infty \{L(x) - x^{1/2}(\zeta(\tfrac{1}{2}))^{-1} - cx^{1/2}\}\, x^{-s-1}dx = \frac{\zeta(2s)}{s\zeta(s)} - \frac{1}{\zeta(\frac{1}{2})(s-\frac{1}{2})} - \frac{c}{s-\frac{1}{2}}$$

and

$$\int_1^\infty \{N(x) + x^{-1/2}(\zeta(\tfrac{1}{2}))^{-1} - cx^{-1/2}\}x^{-s-1}dx = \frac{\zeta(2s+2)}{s\zeta(s+1)} + \frac{1}{\zeta(\frac{1}{2})(s+\frac{1}{2})} - \frac{c}{s+\frac{1}{2}},$$

respectively. Assuming simplicity of the zeros (otherwise the result follows already from [4]), the residue of $\frac{\zeta(2s)}{s\zeta(s)}$ at $s = \rho$ is $\frac{\zeta(2\rho)}{\rho\zeta'(\rho)}$ and equals that of $\frac{\zeta(2s+2)}{s\zeta(s+1)}$ at $s = \rho-1$, because (on the RH!) $|\rho-1| = |\rho|$. The divergence of $\sum \left|\frac{\zeta(2\rho)}{\rho\zeta'(\rho)}\right|$ may be shown using the explicit formula for $L(x)$ analogous to Theorem 14.27 of [16] and the proofs are completed by checking on hand of [7] that $\sum_{n=1}^{13} \left|\frac{\zeta(2\rho)}{\rho\zeta'(\rho)}\right| > .364$, so that, using $N = 16$, $\frac{2N}{N+1} \sum_{n=1}^{13} |\rho_n \zeta'(\rho_n)|^{-1}$ $> .685$. This leads to

$$L(x) \gtrless \left(\frac{1}{\zeta(\frac{1}{2})} \pm c\right) x^{\frac{1}{2}} \qquad \text{and} \qquad N(x) \gtrless \left(-\frac{1}{\zeta(\frac{1}{2})} \pm c\right) x^{-\frac{1}{2}}$$

respectively, with $1/\zeta(\frac{1}{2}) \cong -.68476$ and $c > .685$ and the statements of Theorems 3(ii) and 4(ii) follow.

7. Proof of Theorem 5. Let

$$f_o(x;k,a) = \sum_{\substack{p \le x \\ p^m \equiv a \pmod k}}' m^{-1};$$

then (see [4])

$$(10) \qquad f_o(x;k,a) = \pi(x;k,a) + \frac{\delta(a)n(k)}{2\varphi(k)} \pi(x^{\frac{1}{2}}) + o(x^{\frac{1}{2}}(\log x)^{-m})$$

holds for any fixed m, as $x \to \infty$. Here $\delta(a) = \frac{1}{2}\left(1 + \left(\frac{a}{k}\right)\right)$ and we recall that $n(k)$ is the number of distinct (i.e., incongruent modulo k) solutions of $x^2 \equiv 1 \pmod k$.

In (2) we now set $f(x) = f_o(x;k,a) - f_o(x;k',a')$, with $\varphi(k) = \varphi(k')$, $(a,k) = (a',k') = 1$, take $g(x) = 0$ and

$$u(x) = \sum_j m_{\sigma_j}(a,a') \frac{x^{\sigma_j}}{\log x} + c \frac{x^\theta}{\log x},$$

the sum being extended over all real zeros σ_j of any of the functions $L(s;k,\chi)$ or $L(s;k',\chi')$. Here, exceptionally, we do not start out with the assumption that $\theta = \frac{1}{2}$. With this, (2) becomes

$$\int_2^\infty \left\{f_o(x;k,a) - f_o(x;k',a') - \sum_j m_{\sigma_j}(a,a') \frac{x^{\sigma_j}}{\log x} - c \frac{x^\theta}{\log x}\right\} x^{-s-1} dx$$

$$(11) \qquad = \frac{1}{s\varphi(k)} \left\{\sum_\chi \bar{\chi}(a) \log L(s,\chi) - \sum_{\chi'} \bar{\chi}'(a') \log L(s,\chi')\right\}$$

$$-\sum_j m_{\sigma_j}(a,a')\log\frac{1}{s-\sigma_j} - c\log\frac{1}{s-\theta} + \frac{g(s)}{s}$$

$$= F(s),$$

with $g(s)$ an entire function

With $\sigma_M = \max_{m_\sigma(a,a')\neq 0} \sigma$ and $m_M = m_{\sigma_M}(a,a')$, the inversion of (11) leads to

$$f_o(x;k,a) - f_o(x;k,a') - m_M \frac{x^{\sigma_M}}{\log x}(1+o(1))$$

$$= \frac{1}{2\pi i}\int_{(c)} F(s)x^s ds = h(x),$$

say. Here $F(s)$ is holomorphic for $\sigma > \theta$, so that $h(x) = O(x^{\theta+\epsilon})$, and

(12) $$f_o(x;k,a) - f_o(x;k',a') = m_M \frac{x^{\sigma_M}}{\log x}(1+o(1)).$$

It is clear from its definition that $m_M \neq 0$ and (12) shows that it is real. If we combine (12) with (10) and take into account $\pi(x^{\frac{1}{2}}) \sim 2x^{\frac{1}{2}}/\log x$ and $\sigma_M > \theta \geq \frac{1}{2}$, it follows that

$$D(x) = \pi(x;k,a) - \pi(x;k',a') = m_M \frac{x^{\sigma_M}}{\log x}(1+o(1))$$

has ultimately the sign of m_M. For this conclusion the exact value of θ was not relevant.

The present method allows one to handle also the case $\sigma_M = \theta$ but one is led to case distinctions ($F(s) + c\log\frac{1}{s-\theta}$ holomorphic, or not holomorphic on $\sigma = \theta$; in the latter case, if $F(s)$ has the singularities $r_j\log\{s - (\theta + i\gamma_j)\}^{-1}$ and $r = \sum_j |r_j|$, one may have either $m_M > r$, or $m_M = r$, or $m_M < r$, etc.), the statements are complicated and are suppressed altogether in Theorem 5, because they are almost certainly empty anyhow (most likely $\Sigma = \emptyset$).

We assume now that $\theta > \max(\frac{1}{2}, \sigma_M)$. Two cases have to be considered:

(i) there exist zeros $\rho = \theta + i\gamma$ of the L-functions involved, with $m_\rho(a,a') \neq 0$;

(ii) there are no such zeros.

In the second case, none of our basic theorems applies. In the first case, Theorem A applies, provided that $|c| < |m_{\rho_j}/\rho_j|$ holds for some $\rho_j = \theta + i\gamma_j$ with $m_{\rho_j} \neq 0$. It follows that for some $c > 0$, both inequalities

$$(13) \qquad f_o(x;k,a) - f_o(x;k',a') \gtrless \pm cx^\theta/\log x$$

hold on X-sets. Combining (13) and (10) and taking into account $\theta > \frac{1}{2}$, it follows that both inequalities

$$D(x) \gtrless \pm c'x^\theta/\log x$$

hold on X-sets, for any $c' < c$ and this is the assertion (ii).

If $\sigma_M < \theta = \frac{1}{2}$, then the set of ρ's with $m_\rho \neq 0$ is not empty. Indeed, the opposite assertion would lead to the false conclusion that $f_o(x;k,a) - f_o(x;k',a')$ is a continuous function (see [4], Theorems E and E'). Consequently, by Theorem A, (13) is certain to hold and both inequalities

$$(13') \qquad f_o(x;k,a) - f_o(x;k',a') \gtrless \pm cx^{\frac{1}{2}}/\log x$$

hold on X-sets.

If now, in addition to $\varphi(k) = \varphi(k')$ we also have $n(k)\delta(a) = n(k')\delta(a')$, (i.e., if a and a' are either both quadratic residues or both quadratic non-residues of their respective moduli, and if also the congruence $x^2 \equiv 1$ has the same number of solutions modulo k and modulo k'), then (10) shows that

$$f_o(x;k,a) - f_o(x;k',a') = \pi(x;k,a) - \pi(x;k',a') + o(x^{\frac{1}{2}}(\log x)^{-m}),$$

so that, by (13'), both inequalities

$$D(x) \gtrless \pm cx^{\frac{1}{2}}/\log x$$

hold on X-sets. This is stronger than and implies the first statement of (iii).

If $\delta(a)n(k) \neq \delta(a')n(k')$, then (13′) is insufficient for the desired conclusion. We may use, however, Theorem C with $a_0 = -c$ and $a_j = m_{\rho_j}/\rho_j\varphi(k)$ and conclude that

$$f_0(x;k,a) - f_0(x;k',a') - cx^{\frac{1}{2}}/\log x$$

changes signs on an X-set, whatever the sign of the real coefficient c, provided that

$$(14) \qquad |c| \le \frac{2N}{(N+1)\varphi(k)} \sum_{\gamma \in S} \left|\frac{m_{\rho_j}}{\rho_j}\right|, \qquad \gamma_j = -i(\rho_j - \tfrac{1}{2})$$

with S an N-independent set in $\tilde{\Gamma}$. If C_0 holds, one may select $N = 1$, recall that $\sum |m_\rho/\rho|$ is divergent and conclude that c may be taken arbitrarily large. It follows that

$$\lim_{x\to\infty} {\sup \atop \inf} (\{f_0(x;k,a) - f_0(x;k',a')\} x^{-\frac{1}{2}} \log x) = \pm\infty$$

whence it follows by (10) that also

$$\lim_{x\to\infty} {\sup \atop \inf} (\{\pi(x;k,a) - \pi(x;k',a')\} x^{\frac{1}{2}} \log x) = \pm\infty;$$

this is stronger than, and implies the last assertion of (iii).

More generally, using (10), we have obtained so far that both inequalities

$$D(x) \gtrless \left(\frac{\delta(a')\,n(k') - \delta(a)n(k)}{2\varphi(k)}\right)\frac{x^{\frac{1}{2}}}{\frac{1}{2}\log x} \pm c\,\frac{x^{\frac{1}{2}}}{\log x} + O\left(\frac{x^{\frac{1}{2}}}{\log^2 x}\right)$$

hold on X-sets, provided that (14) holds.

In order to conclude that $D(x) \gtrless 0$ hold on X-sets, we must verify that we may take $|c| > \left|\frac{\delta(a')\,n(k') - \delta(a)n(k)}{\varphi(k)}\right|$, i.e., that

$$\sum_{\gamma \in S} \left|\frac{m_\rho}{\rho}\right| > \frac{N+1}{2N}\,|\delta(a')\,n(k') - \delta(a)n(k)|.$$

For this it is sufficient to have

$$\sum_{\gamma \in S} \left|\frac{m_\rho}{\rho}\right| > \frac{N+1}{2N}\max(n(k), n(k')) = \frac{N+1}{2N} n_o .$$

This finishes the proof of (6), hence that of (iv).

In case $k = k' = p$, we have $n(k) = n(k') = n_o = 2$. If we restrict our attention only to simple zeros (as is obviously sufficient), then $m_\rho = 1$ and (6) becomes (7), and (v) is proven.

It is clear that it is sufficient to consider in (7) only characters for which $\chi(a) \neq \chi(a')$; this excludes, in particular, the principal character $\chi_o(n)$. For the real character,

$$|\chi_{\frac{p-1}{2}}(a) - \chi_{\frac{p-1}{2}}(a')| = 2$$

and, if exactly one of a, a' is a quadratic residue modulo p, with $p \equiv 1 \pmod 4$, then

$$|\chi_{\frac{p-1}{4}}(a) - \chi_{\frac{p-1}{4}}(a')| = \sqrt{2}.$$

More generally, whenever $\chi(a) \neq \chi(a')$, one has $|\chi(a) - \chi(a')| \geq 2\sin\{\pi/(p-1)\}$. Consequently, we try to find a set S of γ's, N-independent in $\tilde{\Gamma}$, so that, with $\rho = \frac{1}{2} + i\gamma$, $\gamma \in S$,

$$(15) \qquad \sum_{\gamma \in S} \left|\frac{\chi(a)-\chi(a')}{\rho}\right|$$

$$\geq 2\sum \frac{1}{|\rho^{(1)}|} + \sqrt{2}\,\frac{1}{|\rho^{(2)}|} + 2\sin\frac{\pi}{p-1}\sum\frac{1}{|\rho^{(3)}|} > 1 + \frac{1}{N};$$

here the first sum is extended over the zeros $\rho^{(1)}$ of $L(s,\chi_{\frac{p-1}{2}})$, the second over the zeros $\rho^{(2)}$ of $L(s,\chi_j)$ with $j = (p-1)/4$, or $j = 3(p-1)/4$ (this sum is empty if $p \not\equiv 1 \pmod 4$)) and the last sum is extended over the zeros $\rho^{(3)}$ of $L(s,\chi_j)$ for $j \neq 0$, $(p-1)/2$, $(p-1)/4$, or $3(p-1)/4$. The existence of such a set S guarantees that both inequalities

$$D(x) = D(x;a,a') \gtrless 0$$

hold on X-sets, for any pair, a, a', with $p \nmid aa'$.

8. Proof of Theorem 5(vi). The case when the quadratic residue is $a = 1$ has been settled by Knapowski and Turán (see [10]; see also Stark [15]). It remains to consider the case of a quadratic residue $a \neq 1$ and a quadratic non-residue a'. In general, the verification of the N-independence by means of (15) is uneconomical. At least for the small primes, a separate reasoning in each case is preferable. The following work is based on the tables [14] of R. Spira, which give the γ's. From each of them $|\rho|$ is found by $|\rho| = \sqrt{\gamma^2 + \frac{1}{4}}$. The computations were performed on the computer CDC 6400 of Temple University, by Professor Hagis, with four decimals. Each "near hit" has been verified by hand to 8 decimals and the corresponding figures are given in Section 9.

8.1. $p = 5$. We use the first two zeros of $L(s,\chi_1)$ and of $L(s,\chi_3)$, and the first three zeros of $L(s,\chi_2)$ (χ_2 = real character). We obtain the sum

$$\sum \left|\frac{\chi(a)-\chi(a')}{\rho}\right|$$

$$> \sqrt{2}\left\{\frac{1}{4.17} + \frac{1}{6.21} + \frac{1}{8.47} + \frac{1}{9.46}\right\} + 2\left\{\frac{1}{6.675} + \frac{1}{9.85} + \frac{1}{11.98}\right\} > 1.5.$$

It follows by (7) that any $N > \frac{1}{.5 + \epsilon}$, i.e., $N = 2$, suffices. The verification that no sum of these $m = 7$ ordinates with coefficients $0, \pm 1, \pm 2$ is again the imaginary part of some zero of some $L(s,\chi)$ with $\chi(n)$ a character modulo 5 requires (see Section 4.3) the consideration of $\frac{1}{2}((2N+1)^m-1) = \frac{1}{2}(5^7-1) = 39,062$ sums. It is comparatively easy to obtain all these sums with the help of a computer, but rather burdensome to check each of them against the (known) γ's of the relevant L-functions. Therefore, and in view of Stark's paper [15] to come out soon, this verification has not been carried out.

8.2. $p = 7$. The residues are $a = 1, 2, 4$; the non-residues are $a' = 3, 5, 6$. The smallest primitive root is $g = 3$. For $a = 2$ and $a' = 3$, or $a' = 6$, one has

$$|\chi_1(a) - \chi_1(a')| = |\chi_5(a) - \chi_5(a')| = 1,$$

$$|\chi_2(a)-\chi_2(a')| = |\chi_4(a) - \chi_4(a')| = \sqrt{3},$$

$$|\chi_3(a)-\chi_3(a')| = 2.$$

We use the zeros of $L(s,\chi_5)$ with $\gamma = 2.509\ldots$; of $L(s,\chi_2)$ with $\gamma = 4.356\ldots$; of $L(s,\chi_4)$ with $\gamma = 6.201\ldots$ and $\gamma = 7.927\ldots$; and of $L(s,\chi_3)$ with $\gamma = 4.475\ldots$ and $\gamma = 6.845\ldots$. This leads to

$$\sum \left|\frac{\chi(a)-\chi(a')}{\rho}\right|$$

$$> 2\left(\frac{11}{4.51} + \frac{1}{6.87}\right) + \sqrt{3}\left(\frac{1}{4.38} + \frac{1}{6.23} + \frac{1}{7.94}\right) + \frac{1}{2.56}$$

$$> 2.01,$$

so that $N = 1$ suffices. The verification that no algebraic sum of these $m = 6$ γ's with coefficients $0, \pm 1$ is again the imaginary part of some zero of some $L(s,\chi)$ with χ mod 5 requires $\frac{1}{2}(3^6-1) = 364$ sums.

For $a = 2$, $a' = 5$,

$$|\chi_1(2) - \chi_1(5)| = |\chi_3(2) - \chi_3(5)| = |\chi_5(2) - \chi_5(5)| = 2,$$

while $\chi_2(2) = \chi_2(5) = \overline{\chi}_4(2) = \overline{\chi}_4(5)$. By using only three of the previous zeros, namely those with $\gamma = 2.509\ldots$, $4.475\ldots$, and $6.845\ldots$, we obtain

$$\sum \left|\frac{\chi(2)-\chi(5)}{\rho}\right| > 2\left\{\frac{1}{2.56} + \frac{1}{4.51} + \frac{1}{6.87}\right\} > 1.51,$$

so that we need $N = 2$; this leads to the consideration of $\frac{1}{2}\{(2\cdot 2+1)^3-1\} = 62$ sums, many of which occured already in the previous computation.

For $a = 4$ and $a' = 3$, $|\chi_j(4) - \chi_j(3)| = 2$ for j odd, $= 0$ otherwise, so that the result follows from that for $a = 2$, $a' = 5$. Finally, for $a = 4$, $a' = 5$, or 6, $|\chi_j(a) - \chi_j(a')|$ takes the same values as for $a = 2$, $a' = 3$, already considered, and this completes the consideration of $p = 7$. Due to the fact that the tables [14] extend only to $|\chi| < 25$, eleven sums still remain to be checked (see Section 9). Professor R. Spira kindly verified that none of these sums corresponds to the imaginary part of a zero of an L-function to the modulus 7.

8.3. $\underline{p = 11}$. The quadratic residues are $a = 1, 3, 4, 5, 9,$ the non-residues are $a' = 2, 6, 7, 8, 10,$ and $g = 2$ is primitive root. From $\chi_7(2) = 3^{7\pi i/5}$ one easily obtains that, for $\chi = \chi_7$,

$$|\chi(4) - \chi(7)| = |\chi(5) - \chi(6)| = |\chi(9) - \chi(2)| = |\chi(3) - \chi(8)| = 2.$$

Using only the first zero of $L(s,\chi_7)$, with $\gamma = 1.231188\ldots$, one obtains $|\rho| < 1.329$, $|\rho|^{-1} > .7524$, so that $\left|\frac{\chi(a) - \chi(a')}{\rho}\right| > 1.504$ and it is sufficient to have $N = 2$, i.e., to verify that $2\gamma = 2.462376\ldots$ is not the imaginary part of any zero of an L-function to the modulus 11, which is in fact the case.

Next, still with $\chi = \chi_7$ and the same zero,

$$\begin{aligned} |\chi(4) - \chi(8)| &= |\chi(4) - \chi(2)| = |\chi(3) - \chi(7)| \\ &= |\chi(3) - \chi(6)| = |\chi(5) - \chi(8)| = |\chi(5) - \chi(10)| \\ &= |\chi(9) - \chi(7)| = |\chi(9) - \chi(10)| \\ &= 2\sin\frac{3\pi}{10} \cong 1.618. \end{aligned}$$

In all these cases $\left|\frac{\chi(a) - \chi(a')}{\rho}\right| > 1.217$. By using (7), we see therefore, that it is sufficient to take $N = 5$ and we immediately verify that $n\gamma = (1.231188\ldots)n$ is not in $\tilde{\Gamma}$ for $n = 2, 3, 4,$ or 5 ($2\gamma = 2.462376\ldots$ and the zero with $\gamma = 2.477243\ldots$ of $L(s,\chi_5)$ are the closest).

In the remaining cases ($a = 4$, $a' = 6$ or 10; $a' = 10$ or 2, etc.)

$$|\chi(a) - \chi(a')| = 2\sin\frac{\pi}{10} \cong .618, \qquad \left|\frac{\chi(a) - \chi(a')}{\rho}\right| > .464.$$

This is less than unity, so that we have to use also another zero. We choose the first zero of the real character χ_5, for which $|\chi_5(a) - \chi(a')| = 2$, $\gamma = 2.477\ldots$, $|\rho| < 2.53$, $2/|\rho| > .7905$, so that $\sum\left|\frac{\chi(a) - \chi(a')}{\rho}\right| > 1.254$. It follows by (7) that $N = 4$ suffices. This requires the consideration of $\frac{1}{2}\{(2\cdot 4 + 1)^2 - 1\} = 40$ sums of these two γ's, with coefficients $0, \pm 1, \pm 2, \pm 3, \pm 4$.

8.4. $\underline{p = 13}$. The smallest primitive root is $g = 2$, the non-principal real character is $\chi_6(n) = \left(\frac{n}{13}\right)$ and $\chi_3(n) = \overline{\chi}_9(n) = \pm i^{\delta(n)+1}$. Consequently, for any $\left(\frac{a}{13}\right) = -\left(\frac{a'}{13}\right) = 1$, $|\chi_6(a) - \chi_6(a')| = 2$, $|\chi_9(a) - \chi_9(a')| = \sqrt{2}$, and, selecting just the first zeros of $L(s,\chi_6)$ and $L(s,\chi_9)$, respectively, with $\gamma = 3.11934...$ and $\gamma = 2.19555...$, we obtain

$$\sum \left|\frac{\chi(a) - \chi(a')}{\rho}\right| > \frac{2}{3.1954} + \frac{\sqrt{2}}{2.2518} > 1.255.$$

By (7) it is sufficient to take $N = 4$ and check $\frac{1}{2}\{(2\cdot 4 + 1)^2 - 1\} = 40$ sums. This finishes the proof for $p = 13$.

8.5. $\underline{p = 17}$. The smallest primitive root is $g = 3$, the real character is $\chi_8(n) = \left(\frac{n}{17}\right)$ and one has $\chi_4(n) = \overline{\chi}_{12}(n) = \pm i^{\delta(n)+1}$. $L(s,\chi_{12})$ has a zero $\rho = \frac{1}{2} + i(1.8945...)$, $|\rho| > 1.96$; $L(s,\chi_8)$ has a zero $\rho = \frac{1}{2} + i(3.728...)$, $|\rho| > 3.76$, so that

$$\sum \left|\frac{\chi(a) - \chi(a')}{\rho}\right| > \frac{2}{3.76} + \frac{\sqrt{2}}{1.96} > 1.254,$$

and $N = 4$ is sufficient. One has to check $\frac{1}{2}\{(2\cdot 4 + 1)^2 - 1\} = 40$ sums and this finishes the proof for $p = 17$.

8.6. $\underline{p = 19}$. The real character is $\chi_9(n) = \left(\frac{n}{19}\right)$ and $L(s,\chi_9)$ has the zero $\rho = \frac{1}{2} + i(1.516...)$, with $|\rho| > 1.5969$. Hence,

$$\left|\frac{\chi(a) - \chi(a')}{\rho}\right| > \frac{2}{1.5969} > \frac{5}{4}$$

and it is sufficient to take $N = 4$, i.e., to verify that 2γ ($= 3.032...$), 3γ ($= 4.548...$), and 4γ ($= 6.064...$) do not occur among the imaginary parts of zeros of $L(s,\chi)$ with $\chi \bmod 19$. (Observe, however, the "near-hits" $\gamma = 3.098...$ of $L(s,\chi_{14})$, $\gamma = 6.144...$ of $L(s,\chi_4)$ and $\gamma = 6.009...$ of $L(s,\chi_7)$). This finishes the proof for $p = 19$.

9. Comments on the computations. The values of the γ's were taken from [14]. Rounded off to 4 decimal places, they were fed into the computer as an ordered set of m elements, together with an integer N. The computer obtained the sums $\sum_{j=-m}^{m} n_j\gamma_j$, $n_j \in Z$, $|n_j| \le N$ and printed out the sums, as well as the values of the corresponding coefficients n_j. As mentioned in Sections 4 and 8, only positive sums are needed. It turned out that it was easier to program the computer to print out all $(2N+1)^m$ sums, regardless of sign. The central sum s_o is zero and the other sums s are printed out so that s and $-s$ are in symmetric positions with respect to $s_o = 0$. In checking the sums it is, therefore, sufficient to consider only the first $\frac{1}{2}\{(2N+1)^m-1\}$ sums preceding $s_o = 0$ and to disregard their sign. There are either 6 summands with coefficients $0, \pm 1$; or 3 summands with coefficients $0, \pm 1, \pm 2$; or 2 summands with coefficients at most ± 4, and the round-off error on each γ is less than 10^{-4}. It follows that the total error on each sum is less than 10^{-3} and can affect at most the third decimal (we disregard cases like a.999..., or b.000..., which in fact do not occur). The resulting sums were checked against the values of all γ's for the functions $L(s,\chi)$ formed with all characters to the respective modulus. Whenever the first two decimals coincided, the corresponding sums were recomputed by hand to eight correct decimal places, to make certain that they do not occur among the γ's. In each case the closest "near hit" was recomputed.

The only difficulties occur for $p = 7$. Eleven of the sums obtained exceed 25, the limit of Spira's tabulation [14] for the γ's. This made it necessary to compute the values of the 5 L-functions modulo 7, for those eleven values $s = \frac{1}{2} + i\gamma$, in order to ascertain that none of them vanishes. Professor R. Spira kindly performed the corresponding verifications.

9.1. Discussion of specific primes.

$p = 7$. The eleven sums exceeding 25 are the following:

32.31566709	27.95926547	27.66120908	26.11443705	25.44989092
29.80629254	27.83992881	27.29691799	25.47017538	25.33055426
				25.15183453

With

$\gamma_1 = 2.50937455,\quad \gamma_2 = 4.35640162,\quad \gamma_3 = 4.47573828,$

$\gamma_4 = 6.20123004,\quad \gamma_5 = 6.84549171,\quad \gamma_6 = 7.92743089,$

and the indicated values for the coefficients n_j, we obtain the following sums $\sum n_j\gamma_j$; we list next to them the values of the γ's that are close. In the last column we indicate (by its number) the character χ, for which $L(s,\chi)$ has as zero $\frac{1}{2} + i\gamma$.

n_1	n_2	n_3	n_4	n_5	n_6	sum	γ	χ	
+1	+1	-1	0	+1	+1	17.16296049	17.16141654	1	
+1	0	+1	+1	+1	0	20.03183458	20.03055898	2	
+1	0	+1	0	+1	0	13.83060454	13.82986789	4	
-1	0	+1	0	0	+1	9.89379463	9.89354379	5	← overall
+1	-1	+1	-1	0	+1	4.35491206	4.35640162	2	closest
+1	-1	0	+1	0	0	4.35420297	4.35640162	2	
-1	+1	0	+1	0	+1	15.97568800	15.93744820	2	
0	0	+1	-1	0	+1	6.20193913	6.20123004	4	

In addition, also the sum $\gamma_2-\gamma_3+\gamma_4+\gamma_5-\gamma_6$, that appears in the print-out as 5.000000 was recomputed to eight decimals and found to be 4.99995220.

With $\gamma_1 = 2.50937455$, $\gamma_2 = 4.47573828$ and $\gamma_3 = 6.84549171$, the sum $-2\gamma_1+2\gamma_2+2\gamma_3 = 17.62371088$ is the closest to a zero, namely of $L(s,\chi_2)$, with $\gamma = 17.61605319$.

<u>p = 11</u>. The sum $3(2.47724371) - 2(1.23118824) = 4.96935465$ is the one that comes closest to a γ, namely to $\gamma = 4.96271516$ of $L(s,\chi_7)$.

<u>p = 13</u>. The sum $4(3.11934147) + 2(2.19555319) = 16.86847226$ is the closest to $\gamma = 16.87785116$ of $L(s,\chi_4)$.

<u>p = 17</u>. The sum $3(3.72814208) + 4(1.89456288) = 18.76267776$ is closest to $\gamma = 18.76358864$ of $L(s,\chi_4)$.

REFERENCES

1. P. T. Bateman, J. W. Brown, R. S. Hall, K. E. Kloss, and Rosemarie M. Stemmler, Linear relations connecting the imaginary parts of the zeros of the zeta function, (to appear).

2. H. Diamond, Two oscillation theorems, these Proceedings, 113-118.

3. D. Davies and C. B. Haselgrove, The evaluation of Dirichlet L-functions, Proc. Roy. Soc. Ser. A 264 (1961), 122-132.

4. E. Grosswald, Oscillation theorems of arithmetical functions, Trans. Amer. Math. Soc. 126 (1967), 1-28.

5. E. Grosswald, On some generalizations of theorems by Landau and Pólya, Israel J. Math. 3 (1965), 211-220.

6. G. H. Hardy and J. E. Littlewood, Contributions to the theory of the Riemann zeta-function and the theory of the distribution of primes, Acta Math. 41 (1918), 119-196.

7. C. B. Haselgrove, A disproof of a conjecture of Pólya, Mathematika 5 (1958), 141-145.

8. C. B. Haselgrove and J. C. P. Miller, Tables of the Riemann Zeta Function, Roy. Soc. Math. Tables 6, Cambridge Univ. Press, Cambridge, 1960.

9. A. E. Ingham, On two conjectures in the theory of numbers, Amer. J. Math. 64 (1942), 313-319.

10. S. Knapowski and P. Turán, Comparative prime number theory, Acta Math. Acad. Sci. Hungar., especially 13 (1962), 229-314, 315-342, 343-364.

11. E. Landau, Über einen Satz von Tschebyscheff, Math. Ann. 61 (1905), 527-550.

12. J. E. Littlewood, Sur la distribution des nombres premiers, C.R. Acad. Sci. Paris 158 (1914), 1869-1872.

13. B. Saffari, Sur les oscillations des fonctions sommatoires des fonctions de Möbius et de Liouville, C. R. Acad. Sci. Paris 271 (1970), 578-580; see also Sur la fausseté de la Conjecture de Mertens, C. R. Acad. Sci. Paris 271 (1970), 1097-2000.

14. R. Spira, Calculation of Dirichlet L-functions, Math. Comp. 23 (1969), 489-498.

15. H. M. Stark, A problem in comparative prime number theory, (to appear).

16. E. C. Titchmarsh, The Theory of the Riemann Zeta Function, The Clarendon Press, Oxford, 1951.

CURVES WITH ABNORMALLY MANY INTEGRAL POINTS*

D. J. Lewis, University of Michigan

1. It is well known that if $f(x)$ is a polynomial with integer coefficients and if for each integer z, $f(z)$ is an m-th power, then $f(x) = h^m(x)$, where $h(x)$ is a polynomial with integer coefficients. There are several ways one could hope to generalize this result. The most obvious is to ask what relation must exist between polynomials f, g with integer coefficients, if $f(\underset{\sim}{Z}) \subset g(\underset{\sim}{Z})$. (Here we let $\underset{\sim}{Z}$ denote the ring of rational integers, $\underset{\sim}{Q}$ the field of rational numbers, $\underset{\sim}{Q}_p$ the field of p-adic integers, $\underset{\sim}{Z}_p$ the ring of p-adic integers, $\underset{\sim}{C}$ the complex field and $f(X)$ the set $\{f(\alpha) \mid \alpha \text{ ranges over } X\}$.) In particular, is $f(x) = g(h(x))$ for some polynomial $h(x)$? The answer is no if we require h to have integer coefficients as is demonstrated by the example: $f = x(x + 1)$, $g = 2x$. However, the answer is yes provided we allow h to have rational coefficients. In that case h is an integer valued function at integer points. Actually, the hypothesis can be relaxed considerably and the same conclusion obtained.

Let $F(x,y)$ be a polynomial with rational coefficients and let $\mathfrak{C} = \mathfrak{C}(F)$ be the affine plane curve defined by F; that is, $\mathfrak{C} = \{(\alpha,\beta) \mid F(\alpha,\beta) = 0, \; \alpha,\beta \in \underset{\sim}{C}\}$. Let $\mathfrak{C}_{\underset{\sim}{Z}}$ be the set of points on $\mathfrak{C}$ with coordinates in $\underset{\sim}{Z}$. Further, let π be the projection from the xy-plane to the x-axis. We have the following general result.

__Theorem__ 1. If $\pi(\mathfrak{C}_{\underset{\sim}{Z}})$ is of positive lower asymptotic density, then there exists a polynomial $h(x)$ in $\underset{\sim}{Q}[x]$ such that $F(x,h(x)) = 0$, or equivalently, $y - h(x)$ is a factor of $F(x,y)$.

__Remark__. The polynomial h in Theorem 1 will be integral valued at a set of integers of positive lower asymptotic density, but need not be integral valued at all integers. For example, h could have the form $\frac{1}{4} x(x + 1)$ which is integral valued only if $x \equiv 0$, $-1 \pmod 4$.

---

*This paper was written while the author was supported in part by a National Science Foundation grant and by the Institute of Science and Technology, University of Michigan.

Proof. Without loss of generality we may assume F has coefficients in $\mathbb{Z}$. Factor F into irreducible factors over $\mathbb{Q}$, which by Gauss's lemma can be assumed to have coefficients in $\mathbb{Z}$. Let G consist of one copy of each irreducible factor g of F which contains y and for which $\pi(\mathfrak{C}_{\mathbb{Z}}(g))$ has positive lower asymptotic density. Clearly G is of positive degree in y and satisfies the hypotheses of the theorem.

If the degree of y in each factor of G is at least 2, by Hilbert's Irreducibility Theorem [2; 3, Chapter 8] there is a set of integers X of lower asymptotic density 0 such that if x_o is in $\mathbb{Z} - X$, each of the factors g of G specialize under the specialization $x \to x_o$ into irreducible polynomials in y of degree at least 2. Hence $\pi[\mathfrak{C}_{\mathbb{Z}}(G)] \subset X$, contrary to the hypothesis. Thus some irreducible factor g of G is linear in y, say $g = A(x)y + B(x)$. Since g is irreducible, $A(x)$ and $B(x)$ have no common factor. Hence there exist polynomials $U(x)$, $V(x)$ in $\mathbb{Z}[x]$ and a non-zero integer c such that $c = A(x)U(x) + B(x)V(x)$. If x is in $\pi[\mathfrak{C}_{\mathbb{Z}}(g)]$ then $A(x)|c$. Since $\pi[\mathfrak{C}_{\mathbb{Z}}(g)]$ is infinite, it follows that $A(x)$ is a non-zero constant, say a. But then $y - a^{-1}B(x)$ is a factor of G, and hence of F, as was to be proved.

As a corollary we have:

Corollary. If $f, g_1, \ldots, g_r$ are in $\mathbb{Q}[x]$ and $f(X) \subset \bigcup_1^r g_i(\mathbb{Z})$, where X is a set of integers of positive lower asymptotic density, then $f = g_j(h(x))$ for some integer j and some polynomial $h(x)$ in $\mathbb{Q}[x]$.

Proof. Let $F(x,y) = \prod [f(x) - g_j(y)]$, and let $\mathfrak{C}$ be the curve defined by $F = 0$. By the hypothesis $\pi[\mathfrak{C}_{\mathbb{Z}}] \supset X$, and hence there is a polynomial $h(x)$ in $\mathbb{Q}[x]$ such that $y - h(x)$ is a component of F. Since $y - h(x)$ is irreducible, it must be a factor of one of the components $f(x) - g_j(y)$; whence the conclusion.

The conclusion of Theorem 1 holds even if $\pi[\mathfrak{C}_{\mathbb{Z}}]$ has lower asymptotic density 0, provided $\pi[\mathfrak{C}_{\mathbb{Z}}]$ is distributed with some "regularity" in $\mathbb{Z}$. In particular, Davenport, Lewis and Schinzel [1] have proved:

<u>Theorem</u> 2. If $\pi[\mathfrak{C}_{\underset{\sim}{Z}}]$ meets each arithmetic progression, then there exists a polynomial $h(x)$ in $\underset{\sim}{Q}[x]$ such that $y - h(x)$ is a factor of F.

The proof of Theorems 1 and 2 have the same general format--in each case one proves the existence of a factor of F which is linear in y and which vanishes at an infinity of integer points. However, in the case of Theorem 2, the proof of this fact requires the more sophisticated concepts and results of class field theory. We cannot assert that the polynomial $h(x)$ in Theorem 2 is integer valued, as demonstrated by the example

$$F(x,y) = (3y - x - x^2)(3y + 1 - x^2).$$

Here $\pi[\mathfrak{C}_{\underset{\sim}{Z}}] = \underset{\sim}{Z}$, but this is not so for either component.

It should be observed that while the hypothesis for Theorem 2 is in some sense weaker than that for Theorem 1, still the assumption that $\pi(\mathfrak{C}_{\underset{\sim}{Z}})$ meets each arithmetic progression implies that it meets each arithmetic progression infinitely often, and hence is quite strong in its own right.

2. Theorems 1 and 2 have a very striking conclusion: namely that $\mathfrak{C}$ has a component in which the projection map is one-half of a birational map of the component onto the line. One cannot expect to get such a strong conclusion without appropriately strong hypotheses. For instance, the curve $\mathfrak{C}$: $x^2 = y^3$ has infinitely many integral points, but $\mathfrak{C}$ is irreducible and hence $\mathfrak{C}$ does not satisfy the conclusion of Theorems 1 and 2. In this case, $\pi(\mathfrak{C}_{\underset{\sim}{Z}})$ is the set of cubes and that set clearly misses several arithmetic progressions modulo 7. On the other hand, we do have the following result:

<u>Theorem</u> 3. If $\mathfrak{C}_{\underset{\sim}{Z}}$ is an infinite set, then $\mathfrak{C}$ has an irreducible component $\mathfrak{C}^*$ of genus 0 defined over $\underset{\sim}{Q}$ and there exist rational functions $P(t)/D(t)$, $R(t)/D(t)$ from $\underset{\sim}{Q}(t)$ such that $\mathfrak{C}^*$ contains $\mathrm{loc}_{\underset{\sim}{Q}}(P(t)/D(t), R(t)/D(t))$, whence $F(P(t)/D(t), R(t)/D(t)) = 0$ identically. Furthermore, the polynomial $D(t)$ is either a power of a linear polynomial or a power of a quadratic polynomial.

Proof. This is but an elementary deduction from theorems of Siegel, see [5, 7, 3], regarding absolutely irreducible curves defined over $\mathbb{Q}$ containing an infinity of integral points. To apply that theorem we need only show that F has a factor with coefficients in $\mathbb{Q}$ which is irreducible in $\mathbb{C}[x,y]$ and which vanishes for an infinity of integral points. Suppose such were not the case. Then F is the product of a factor which vanishes only at finitely many integral points and finitely many other factors each of which vanish at infinitely many integral points, each of which is irreducible in $\mathbb{C}[x,y]$ and each of which has some ratio of its coefficients outside $\mathbb{Q}$. Without loss of generality, we can further assume that each factor has 1 amongst its coefficients. Let g be one of these factors and let $g = g^{(1)}, g^{(2)}, \ldots, g^{(m)}$ be the factors of F which are conjugate to g. By assumption $m \geq 2$. Clearly, $\mathfrak{C}_{\mathbb{Z}}(g) = \mathfrak{C}_{\mathbb{Z}}(g^i)$, $i = 1, \ldots, m$. Now

$$G = g^{(1)} \cdots g^{(m)} \quad \text{and} \quad H = g^{(1)} + \cdots + g^{(m)}$$

define curves having an infinity of common points and so have a common factor. Since the $g^{(i)}$ are irreducible over $\mathbb{C}$, this common factor must be one of the $g^{(i)}$. But G and H have coefficients in $\mathbb{Q}$ and hence by conjugation each of the $g^{(i)}$ are a common factor. Since $m \geq 2$, by comparison of degrees we see that H is the zero polynomial. But this is impossible, since the coefficients of H are traces of the coefficients of g and g has 1 as a coefficient. Hence F must have an irreducible factor g with coefficients in $\mathbb{Q}$ and having an infinity of integral points, and the theorem now follows from that of Siegel when applied to g.

3. We next prove a generalization of Theorem 2; similar generalizations exist for our other theorems.

Theorem 4. Let $\mathfrak{C}$ be a curve defined over $\mathbb{Q}$ lying in n-dimensional space $\mathbb{A}^n$. Choose a fixed coordinization of $\mathbb{A}^n$ and relative to that coordination, let $\mathfrak{C}_{\mathbb{Z}}$ denote the points on $\mathfrak{C}$ with rational integers as coordinates. Let π_1 be the projection of $\mathbb{A}^n$ onto the x_1-axis. If $\pi_1(\mathfrak{C}_{\mathbb{Z}})$ meets each arithmetic progression, then $\mathfrak{C}$ contains a component of the form

$$\mathrm{loc}_{\underset{\sim}{Q}}(t,\ h_2(t),\ \ldots,\ h_n(t)),$$

where $h_i(t)$ are polynomials over $\underset{\sim}{Q}$.

<u>Proof</u>. Let $\mathfrak{C}_1, \ldots, \mathfrak{C}_m$ be those components of $\mathfrak{C}$ with generic points $\underset{\sim}{\xi}_1 = (\xi_{11}, \ldots, \xi_{1n}), \ldots, \underset{\sim}{\xi}_m = (\xi_{m1}, \ldots, \xi_{mn})$, respectively, such that $\pi_1(\mathfrak{C}_{i\underset{\sim}{Z}})$ is an infinite set. Then the $\pi_1(\underset{\sim}{\xi}_i) = \xi_{i1}$ are transcendental over $\underset{\sim}{Q}$. Since $\pi_1(\mathfrak{C}_{\underset{\sim}{Z}})$ is infinite, such components must exist. Let $\mathfrak{C}^*$ be the union of the components $\mathfrak{C}_1, \ldots, \mathfrak{C}_m$. Then $\pi_1(\mathfrak{C}_{\underset{\sim}{Z}}) - \pi_1(\mathfrak{C}^*_{\underset{\sim}{Z}})$ is a finite set. Furthermore, if $\mathfrak{C}'$ is a component of $\mathfrak{C}$ conjugate to one of the $\mathfrak{C}_i$, then $\mathfrak{C}'$ is one of the $\mathfrak{C}_i$. Hence $\mathfrak{C}^*$ is defined over $\underset{\sim}{Q}$. Our construction is such that $\mathfrak{C}^*$ satisfies the hypotheses of the theorem.

Now for each i, the field $\underset{\sim}{Q}(\underset{\sim}{\xi}_i) = \underset{\sim}{Q}(\xi_{i1}, \ldots, \xi_{in})$ is a function field over $\underset{\sim}{Q}$ of transcendency degree 1 and is a finite algebraic extension of $\underset{\sim}{Q}(\xi_{i1})$. Hence, by the usual techniques, we can find integers $a_2, \ldots, a_n$ such that

$$\underset{\sim}{Q}(\underset{\sim}{\xi}_i) = \underset{\sim}{Q}(\xi_{i1}, \eta_i), \quad (i = 1, \ldots, m),$$

where

$$\eta_i = a_2 \xi_{i2} + \cdots + a_r \xi_{in}, \quad (i = 1, \ldots, m),$$

and

$$\xi_{ij} = \frac{p_{ij}(\xi_{i1}, \eta_i)}{q_i(\xi_{i1}, \eta_i)}, \quad (i = 1, \ldots, m;\ j = 2, \ldots, n).$$

Let θ be the map of $\underset{\sim}{A}^n$ into $\underset{\sim}{A}^2$ given by

$$\theta(x_1, \ldots, x_n) = (x_1, a_2x_2 + \cdots + a_nx_n).$$

Then θ is one-half of a birational mapping of $\mathfrak{C}$ into A^2. Since the ξ_{i1} $(i = 1, \ldots, m)$, are transcendental over $\underset{\sim}{Q}$, it follows that $\theta(\mathfrak{C}_1), \ldots, \theta(\mathfrak{C}_m)$ are irreducible curves in $\underset{\sim}{A}^2$ and, moreover, $\theta(\mathfrak{C}^*)$ is a curve in $\underset{\sim}{A}^2$ defined over $\underset{\sim}{Q}$. Clearly, θ carries integral points into integral points. Hence

$$\pi(\theta(\mathfrak{C}^*)_{\underset{\sim}{Z}}) \supset \pi_1(\mathfrak{C}^*_{\underset{\sim}{Z}}).$$

Thus we have shown that $\theta(\mathfrak{C}^*)$ satisfies the hypotheses of Theorem 2, and hence it follows from that theorem that $\theta(\mathfrak{C}^*)$ contains a component Γ of the form: $\mathrm{loc}(t, h(t))$, where h is a polynomial over $\underset{\sim}{Q}$.

Let $\mathfrak{C}_k$ be a component of $\mathfrak{C}^*$ such that $\theta(\mathfrak{C}_k) = \Gamma$. Then $\mathfrak{C}_j = \theta^{-1}(\Gamma)$ and so

$$\mathfrak{C}_k = \mathrm{loc}_{\underset{\sim}{Q}}\left(t, \frac{p_{k2}(t,h(t))}{q_k(t,h(t))}, \ldots, \frac{p_{kn}(t,h(t))}{q_k(t,h(t))}\right),$$

or

$$\mathfrak{C}_k = \mathrm{loc}_{\underset{\sim}{Q}}\left(t, \frac{h_2(t)}{q(t)}, \ldots, \frac{h_n(t)}{q(t)}\right),$$

where $q(t), h_2(t), \ldots, h_n(t)$ are polynomials over $\underset{\sim}{Z}$ without a nonconstant common factor. Then there exist polynomials $v(t), v_2(t), \ldots, v_n(t)$ in $\underset{\sim}{Z}[t]$ and a non-zero integer c such that

$$q(t)v(t) + h_2(t)v_2(t) + \cdots + h_n(t)v_n(t) = c.$$

By construction $\pi_1(\mathfrak{C}_{k\underset{\sim}{Z}})$ is infinite, and hence for infinitely many integers α we see that $h_i(\alpha) = (\text{integer})q(\alpha)$, $(i = 2, \ldots, m)$. It then follows that $q(\alpha) | c$ for infinitely many integers α; whence $q(t)$ is a constant. This completes the proof of the theorem.

4. Our condition that $\pi_1(\mathfrak{C}_{\underset{\sim}{Z}})$ meet each arithmetic progression may, at first glance, seem to be on the artificial side. In the study of diophantine problems one often finds it advantageous to study local conditions. Our condition implies very simple local conditions.

Suppose a coordinization for $\underset{\sim}{A}^n$ has been chosen and is fixed throughout our discussion. Let $\mathfrak{C}$ be a curve in $\underset{\sim}{A}^n$ defined over $\underset{\sim}{Q}$. Let $\mathfrak{C}_p$ denote the set of points on $\mathfrak{C}$ all of whose coordinates lie in the ring of p-adic integers $\underset{\sim}{Z}_p$.

<u>Theorem</u> 5. If $\pi_1(\mathfrak{C}_{\underset{\sim}{Z}})$ meets each arithmetic progression, then $\pi_1(\mathfrak{C}_p) = \underset{\sim}{Z}_p$ for all primes p.

<u>Proof</u>. Let p be a fixed prime and let α be an element of $\underset{\sim}{Z}_p$. Then for each positive integer m, $\alpha = a_m + p^m x_m$, where a_m is in $\underset{\sim}{Z}$ and x_m is in $\underset{\sim}{Z}_p$. By one hypothesis, there exist integral points $\underset{\sim}{Y}_m = (Y_{m1},\ldots,Y_{mn})$ on $\mathfrak{C}$ with $Y_{m1} \equiv a_m \equiv \alpha \pmod{p^m}$. Since $\underset{\sim}{Z} \subset \underset{\sim}{Z}_p$, and $\underset{\sim}{Z}_p$ is compact, the sequence $\underset{\sim}{Y}_1, \ldots, \underset{\sim}{Y}_m, \ldots$ has an accumulation point $\underset{\sim}{Y}$ in the affine space $\underset{\sim}{Z}_p^n$. Since $\mathfrak{C}$ is defined by polynomials with coefficients in $\underset{\sim}{Q}$, it follows, by continuity, that $\underset{\sim}{Y}$ is on $\mathfrak{C}$. Since $Y_{m1} \equiv \alpha \pmod{p^m}$, we see that α is the p-adic limit of the Y_{m1}. Hence $\pi_1(\underset{\sim}{Y}) = \alpha$. Since α was an arbitrary element of $\underset{\sim}{Z}_p$, we have $\pi_1(\mathfrak{C}_p) = \underset{\sim}{Z}_p$. This completes the proof of the theorem.

The converse of Theorem 5 is not true. Indeed, we can have $\pi_1(\mathfrak{C}_p) = \underset{\sim}{Z}_p$ for all p, and have $\mathfrak{C}_{\underset{\sim}{Z}}$ empty; for example, consider

$$F = (1 + 3y + 3x^2)(1 + 5y + 5x^3),$$

or

$$F = (1 + x^2y)(1 + (x + 1)^2y).$$

Also, the condition $\pi_1(\mathfrak{C}_p) = \underset{\sim}{Z}_p$ for all p is not sufficient to imply the conclusion to Theorem 2 (hence also to Theorem 4). For the polynomial

$$F = (6x^2 - y^2)(10x^2 - y^2)(15x^2 - y^2)(241x^2 - y^2)$$

has $\pi_1(\mathfrak{C}_p) = \underset{\sim}{Z}_p$ for all p, and F has no linear factors over $\underset{\sim}{Q}$. In this case, there is but one point in $\mathfrak{C}_{\underset{\sim}{Z}}$, namely the origin.

It would be of interest to know if the hypothesis $\pi_1(\mathfrak{C}_p) = \underset{\sim}{Z}_p$, for all p (or all large p), necessarily implies that $\mathfrak{C}$ contains a component defined over $\underset{\sim}{Q}_p$ of the form $\mathrm{loc}_{\underset{\sim}{Q}_p}(t, h_2(t), \ldots, h_n(t))$, where the h_i are polynomials over $\underset{\sim}{Q}_p$. We note that if $n = 2$ and $\mathfrak{C}$ contains a component $\mathrm{loc}_F(t, h_2(t))$, where F is any field of characteristic 0 and $h_2(t)$ is in $F[t]$, then the coefficients of h_2 lie in an algebraic extension of $\underset{\sim}{Q}$.

<u>Theorem</u> 6. Let $\mathfrak{C}$ be an absolutely irreducible curve in $\underset{\sim}{A}_n$ defined over $\underset{\sim}{Q}$ and suppose that $\mathfrak{C}_{\underset{\sim}{Z}}$ is an infinite set and that $\pi_1(\mathfrak{C}_p) = \underset{\sim}{Z}_p$ for all large p. Then $\mathfrak{C} = \mathrm{loc}_{\underset{\sim}{Q}}(t, h_2(t), \ldots, h_n(t))$, where the h_i are polynomials over $\underset{\sim}{Q}$.

<u>Proof</u>. Since $\mathfrak{C}$ is irreducible and $\mathfrak{C}_{\underset{\sim}{Z}}$ is an infinite set, it follows from a theorem of Siegel [3, 5] that

$$\mathfrak{C} = \mathrm{loc}_{\underset{\sim}{Q}}\left(\frac{P_1(t)}{D(t)}, \frac{P_2(t)}{D(t)}, \ldots, \frac{P_n(t)}{D(t)}\right),$$

where $P_1, \ldots, P_n$ and D are polynomials over $\underset{\sim}{Z}$. Since $\pi_1(\mathfrak{C}_p) = \underset{\sim}{Z}_p$ for some p, it follows that P_1 is not a scalar multiple of D. Suppose either $\deg P_1 > 1$ or $\deg D > 1$. Then $P_1(t) - xD(t)$ is irreducible in $\underset{\sim}{Q}[x,t]$ and consequently, by Hilbert's irreducibility theorem, there exist integers a such that $H(t) = P_1(t) - aD(t)$ is irreducible of degree at least 2. But then there exists an infinity of primes p such that $H(t) \equiv 0 \pmod p$ has no solution modulo p and consequently $H(t)$ has no zero in $\underset{\sim}{Z}_p$. But then a is not in $\pi_1(\mathfrak{C}_p)$, contrary to one hypothesis. Hence we can suppose both P_1 and D are of degree at most 1.

Now suppose that D is of degree 1. We can write

$$\frac{P_1(t)}{D(t)} = \frac{\alpha t + \beta}{\gamma t + \delta},$$

where $\alpha, \beta, \gamma, \delta$ are in $\underset{\sim}{Z}$ and $\gamma \neq 0$. Furthermore, since $P_1(t)/D(t)$ is not a constant, we have $\alpha\delta - \beta\gamma \neq 0$. Hence there exist integers a, b, c, d such that

$$x = \frac{\alpha t + \beta}{\gamma t + \delta}, \qquad t = \frac{ax + b}{cx + d}$$

are inverse fractional linear mappings. Thus $ad - bc \neq 0$ and we can write

$$\mathfrak{C} = \mathrm{loc}_{\underset{\sim}{Q}}\left(x, \frac{R_2(x)}{E(x)}, \ldots, \frac{R_n(x)}{E(x)}\right).$$

where

$$E(x) = (Mx + N)(cx + d)^{m-1}, \quad M, N \text{ not both } 0,$$

$$R_i(x) = (cx + d)^{m-1} P_i\left(\frac{ax + b}{cx + d}\right)$$

$$= e_i(ax + b)^{m_i}(cx + d)^{m-m_i} + (cx + d)V_i(x)$$

$$(i = 2, \ldots, n),$$

and $m = \max\{m_i\}$, where $m_i = \deg P_i$.

Without loss of generality, we assume $m = m_2$. Suppose $m \geq 2$. Since, for all large p, we have $\pi_1(\mathfrak{C}_p) = \underset{\sim}{Z}_p$, we see that for p sufficiently large,

$$\text{ord}_p R_2(x) \geq \text{ord}_p E(x),$$

for all x in $\underset{\sim}{Z}_p$. Let q be among these primes and suppose $q > c$. Then

$$\text{ord}_q R_2(-d/c) \geq \text{ord}_q E(-d/c) = \infty.$$

But

$$R_2(x) = e_2(ax + b)^m + (cx + d)V_2(x),$$

whence $a(-d/c) + b = 0$ contrary to a result derived earlier. Hence if $\deg D = 1$, we must have $m \leq 1$. If now $M \neq 0$, then the assumption that $R_i(x)/E(x)$ is p-adic integral for all x in $\underset{\sim}{Z}_p$ implies that the R_i are scalar multiples of E, whence

$$\mathfrak{C} = \text{loc}_{\underset{\sim}{Q}}(x, e_2, \ldots, e_n).$$

If $m < 1$ and $M = 0$, $\mathfrak{C}$ again has the desired form.

On the other hand, if D is a nonzero constant, we must have $\deg P_1 = 1$, whence $\delta x_1 = (\alpha t + \beta)$ and we can express the $P_i(t)$ as a polynomial in x_1 with coefficients in $\underset{\sim}{Q}$; that is, the conclusion of the theorem holds.

The hypotheses of Theorem 6 cannot be appreciably weakened as the curves defined by the polynomial equations:

$$F_1 = x^2 - y^3 = 0,$$

$$F_2 = (6x^2 - y^2)(10x^2 - y^2)(15x^2 - y^2)(241x^2 - y^2) = 0,$$

$$F_3 = F_1F_2 = 0,$$

demonstrate.

5. Recall that $\underset{\sim}{Z}^k$, the set of k-th powers of integers is a multiplicative semigroup. Another generalization of the hypothesis $f(\underset{\sim}{Z}) \subset \underset{\sim}{Z}^k$ is to inquire what the condition $f(\underset{\sim}{Z}) \subset S$, S a multiplicative semigroup of $\underset{\sim}{Z}$, implies regarding the shape of the polynomial f. Obviously, one cannot expect to obtain any information regarding f if S has positive lower asymptotic density. In a series of papers, Davenport, Schinzel, Zassenhaus and Lewis, collectively and individually [1, 4, 6], discussed the case where $S = \underset{\sim}{Z} \cap N_K$, where N_K is the set of rationals which are norms of elements in an algebraic number field K. Here we shall discuss the case where $S = \underset{\sim}{Z}^{e_1} \cdots \underset{\sim}{Z}^{e_m}$. We prove

<u>Theorem</u> 7. Let $f(x)$ be a polynomial with coefficients in $\underset{\sim}{Z}$. If $f(\underset{\sim}{Z}) \subset \underset{\sim}{Z}^{e_1} \cdots \underset{\sim}{Z}^{e_n}$, then $f = g_1^{e_1} \cdots g_n^{e_n}$, where the g_i are polynomials over $\underset{\sim}{Z}$.

<u>Proof</u>. If $(e_1, \ldots, e_n) = e$, then $f(\underset{\sim}{Z}) \subset \underset{\sim}{Z}^{e_1} \cdots \underset{\sim}{Z}^{e_n} \subset \underset{\sim}{Z}^e$ and by Theorem 1, $f = g^e$ where g is a polynomial over $\underset{\sim}{Q}$. By Gauss's lemma, it follows that $f = ch^e$, where c and the coefficients of h are in $\underset{\sim}{Z}$. On evaluating the equality at a point other than a zero of $f(x)$, we see that c is an e-th power. Hence $f = F^e$, where F has coefficients in $\underset{\sim}{Z}$.

Now let $e_i = es_i$. For each integer z we have $f(z) = F^e(z) = w^e$, where w is in $\underset{\sim}{Z}^{s_1} \cdots \underset{\sim}{Z}^{s_n}$. It follows that $F(\underset{\sim}{Z}) \subset \underset{\sim}{Z}^{s_1} \cdots \underset{\sim}{Z}^{s_n}$ if e is odd, and $F(\underset{\sim}{Z}) \subset \pm \underset{\sim}{Z}^{s_1} \cdots \underset{\sim}{Z}^{s_n}$ if e is even. Since the sign of $F(z)$ is the same for all large z, on choosing the sign of F appropriately we can conclude, for some N,

$$F(\underset{\sim}{Z}(N)) \subset \underset{\sim}{Z}^{s_1} \cdots \underset{\sim}{Z}^{s_n},$$

where $\underset{\sim}{Z}(N)$ is the set of integers greater than N.

Hence, we may now suppose (without loss of generality)

$$f(\underset{\sim}{Z}(N)) \subset \underset{\sim}{Z}^{e_1} \cdots \underset{\sim}{Z}^{e_n},$$

where $(e_1, \ldots, e_n) = 1$ and N is sufficiently large. We now factor f into a product of polynomials irreducible over $\underset{\sim}{Q}$. By Gauss's lemma we can assume they have coefficients in $\underset{\sim}{Z}$. Say

$$f(x) = ch_1^{m_1} \cdots h_r^{m_r},$$

where c is a nonzero integer. If each m_i is in $Z(0)e_1 + \cdots + Z(0)e_n = \mathfrak{m}$, then $f = cg_1^{e_1} \cdots g_n^{e_n}$. Furthermore, since c is in $\underset{\sim}{Q}^{e_1} \cdots \underset{\sim}{Q}^{e_n}$ if $p^a \| c$ then a is in $\mathfrak{m}$. Hence in this case we do have the conclusion holding.

Next suppose

$$f = h^m H,$$

where m is not in $\mathfrak{m}$ and $h^m = h_i^{m_i}$ for some i. Since h, H are relatively prime, there exist polynomials $A(x)$, $B(x)$ and an integer d such that

$$hA + HB = d.$$

There are infinitely many primes p such that $h(x) \equiv 0 \pmod p$ is soluble. Choose such a prime p which exceeds d and the discriminant of $h(x)$, and let α be an integer exceeding N such that $h(\alpha) \equiv 0 \pmod p$. Then $H(\alpha)$ is not divisible by p and since the exact power of p dividing $f(\alpha)$ lies in $\mathfrak{m}$ we conclude that $h(\alpha) \equiv 0 \pmod{p^2}$. Now we also have $h(\alpha + p) \equiv 0 \pmod p$ and by the same reasoning can conclude $h(\alpha + p) \equiv 0 \pmod{p^2}$. But

$$h(\alpha + p) \equiv h(\alpha) + ph'(\alpha) \pmod{p^2},$$

hence $h'(\alpha) \equiv 0 \pmod p$; but this is contrary to our choice of p to exceed the discriminant of $h(x)$. Thus we see that the assumption that m is not in $\mathfrak{M}$ leads to nonsense, and hence the conclusion of the theorem follows.

REFERENCES

1. H. Davenport, D. J. Lewis and A. Schinzel, Polynomials of certain special types, Acta Arith. 9 (1964), 108-116.

2. D. Hilbert, Über die Irreduzibiltat ganzer rationaler Funktionen mit ganzzahligen Koeffizienten, Ges. Abhandlungen II, 264-286.

3. S. Lang, Diophantine Geometry, Addison-Wesley, New York, 1962.

4. D. J. Lewis, A. Schinzel and H. Zassenhaus, An extension of the theorem of Bauer and polynomials of certain special types, Acta Arith. 11 (1966), 346-352.

5. C. L. Siegel, Über einige Anwendungen Diophantischer Approximationen, Abh. Preuss. Akad. Kl. Phys. Math. (1929), 41-69.

6. A. Schinzel, On a theorem of Bauer and some of its applications, Acta Arith. 11 (1966), 333-344.

7. Th. Skolem, Diophantische Gleichungen, Ergbnisse der Mathematik 4, Berlin, 1938.

ON CERTAIN ARITHMETICAL SUMS

K. Nageswara Rao, North Dakota State University

1. Introduction. In [7] Menon has established the very elegant result that

$$(1.1) \qquad \sum_{a \pmod n} (a-1,n) = \varphi(n)d(n),$$

where the summation is over all integers a in a reduced residue system modulo the natural number n; here $\varphi(n)$ is Euler's totient and $d(n)$ is the number of positive divisors of n. Menon has given three different proofs of (1.1). By using an entirely different method based on Cauchy composition and finite Fourier representations we obtain generalizations of (1.1) in two different directions.

We take this opportunity to thank Dr. P. Kesava Menon for kindly providing us with a copy of his paper prior to publication.

2. Main results. Let k be a fixed but arbitrary positive integer. A k-vector mod n is an ordered set of integers $[a_1,\dots,a_k]$ where two vectors are considered the same if the corresponding components are congruent mod n. A k-vector $[a_1,\dots,a_k]$ is said to be prime to n if $(a_1,\dots,a_k,n) = 1$. Let $J_k(n)$ be the Jordan totient, namely the number of k-vectors mod n, prime to n. We now state our

Theorem 1. Let $[s_1,\dots,s_k]$ be a (fixed) k-vector mod n, which is prime to n. Then

$$(2.1) \qquad \sum (a_1-s_1,\dots,a_k-s_k,n)^k = J_k(n)d(n),$$

where the sum is taken over all those vectors $[a_1,\dots,a_k]$ mod n that are prime to n.

Denote by $(a,n^k)_k$ the greatest of those common divisors of a and n^k which is a k-th power. Let $\varphi_k(n)$ be Cohen's totient, namely, the number of integers a in a complete residue system (C. R. S.) mod n^k which are k-prime to n^k, i.e., which are such that $(a,n^k)_k = 1$. We then have

Theorem 2. Let s be (fixed) integer k-prime to n^k. Then

$$(2.2) \qquad \sum (a-s,n^k)_k = \varphi_k(n)d(n),$$

where the summation is over all those integers a in a C. R. S. mod n^k that are k-prime to n^k.

Clearly (2.1) and (2.2) reduce to (1.1) in case $k = 1$ and the s are equal to unity. We also observe that in view of the known equality (Cohen [4]) $J_k = \varphi_k$, the sums in (2.1) and (2.2) have equal values. To make the paper easy to follow we introduce some terminology and state some known results.

3. Preliminaries. An arithmetic function $f(m,n)$ is said to be even (mod n) if $f(m,n) = f(m',n)$ whenever $(m,n) = (m',n)$. An arithmetic function $f(m_1,\ldots,m_k,n)$ is said to be totally even (mod n) (Cohen [5]) if there exists an arithmetic function $F(m,n)$ which is even (mod n) and is such that $f(m_1,\ldots,m_k,n) = F(m,n)$ where $m = (m_1,\ldots,m_k)$. Every totally even function $f(m_1,\ldots,m_k,n)$ (mod n), has the unique representation (Cohen [5])

$$(3.1) \qquad f(m_1,\ldots,m_k,n) = \sum_{d|n} \alpha_d c(m_1,\ldots,m_k,d),$$

where the summation runs over all the positive divisors of n and

$$(3.2) \qquad \alpha_d = \frac{1}{n^k} \sum_{\delta|n} F\left(\frac{n}{\delta},n\right) c^{(k)}\left(\frac{n}{d},\delta\right).$$

Here and elsewhere in the paper $c(m_1,\ldots,m_k,n)$ and $c^{(k)}(m,n)$ are extensions, due to Cohen [5], of Ramanujan's sum and are defined thus:

$$(3.3) \qquad c(m_1,\ldots,m_k,n) = \sum \exp 2\pi i(m_1x_1+\cdots+m_kx_k)/n,$$

where the summation is over all k-vectors $[x_1,\ldots,x_k]$ (mod n) which are prime to n; and

$$(3.4) \qquad c^{(k)}(m,n) = c(m,m,\ldots,m,n) \quad (m,\ k \text{ times}).$$

The Cauchy composite $h = f \cdot g$ of two totally even functions f and g (mod n) is defined by setting

$$(3.5) \qquad h(m_1,\ldots,m_k,n) = \sum f(a_1,\ldots,a_k,n)g(b_1,\ldots,b_k,n)$$

where $m_i \equiv a_i + b_i \pmod n$ and a_i, b_i are allowed to vary over a C. R. S. (mod n). If f has the representation (3.1) and

$$(3.6) \qquad g(m_1,\ldots,m_k,n) = \sum \beta_d \; c(m_1,\ldots,m_k,d),$$

where the sum runs over all the positive divisors d of n, then the Cauchy composite $h = f \cdot g$ has the representation

$$(3.7) \qquad h(m_1,\ldots,m_k,n) = n^k \sum_{d|n} \alpha_d \beta_d \; c(m_1,\ldots,m_k,d).$$

An arithmetic function $f_k(m,n)$ is said to be k-<u>even</u> (mod n^k) if

$$f_k(m,n) = f_k(m',n)$$

whenever

$$(m,n^k)_k = (m',n^k)_k.$$

Every k-even function $f_k(m,n)$ (mod n^k) has the unique representation (see McCarthy [8])

$$(3.8) \qquad f_k(m,n) = \sum_{d|n} \alpha_d \; c_k(m,d),$$

the summation being over all those positive divisors d of n. Here

$$(3.9) \qquad \alpha_d = \frac{1}{n^k} \sum_{\delta|n} f_k(\delta^k,n)\, c_k\!\left(\frac{n}{d^k},\frac{n}{\delta}\right)$$

where $c_k(m,n)$ denotes the following extension of Ramanujan's sum (cf. Cohen [2]):

$$(3.10) \qquad c_k(m,n) = \sum \exp\,(2\pi imx/n^k),$$

the summation running over all those integers $x \pmod{n^k}$ that are k-prime to n^k.

The Cauchy composite $h_k \equiv f_k \cdot g_k$ of two k-even functions f_k, g_k is defined thus:

$$h_k(m,n) = \sum_{m \equiv a+b \pmod{n^k}} f_k(a,n) g_k(b,n), \tag{3.11}$$

where a and b are allowed to vary over a C. R. S. (mod n^k). As before, if f_k has the representation (3.8) and

$$g_k(m,n) = \sum_{d|n} \beta_d \, c_k(m,d),$$

then

$$h_k(m,n) = n^k \sum_{d|n} \alpha_d \beta_d \, c_k(m,d). \tag{3.12}$$

4. Lemmata and Proof of Theorem 1.

Lemma 1. If $d|m$, then

$$\sum_{D|m} D^k c^{(k)}\left(\frac{m}{d}, \frac{m}{D}\right) = \sum_{\delta | \frac{m}{d}} J_k\left(\frac{m}{\delta}\right) \delta^k.$$

Proof. It is known (Cohen [5, (2.10)]) that

$$c^{(k)}(m,n) = \sum_{d|(m,n)} d^k \mu\left(\frac{n}{d}\right). \tag{4.1}$$

Hence

$$\sum_{D|m} D^k c^{(k)}\left(\frac{m}{d}, \frac{m}{D}\right) = \sum_{D|m} D^k \sum_{\delta \left| \left(\frac{m}{d}, \frac{m}{D}\right)\right.} \delta^k \mu\left(\frac{m}{D\delta}\right)$$

$$= \sum_{\delta | \frac{m}{d}} \delta^k \sum_{D | \frac{m}{\delta}} D^k \mu\left(\frac{m}{D\delta}\right),$$

the second equality resulting by a change in summation orders. Since it is well known (Cohen [5, (2.1)]) that

(4.2) $$\sum_{D|m} D^k \mu\left(\frac{m}{D}\right) = J_k(m),$$

the lemma is now proved.

Lemma 2.

$$\frac{m^k}{J_k(m)} = \sum_{d|m} \frac{\mu^2(d)}{J_k(d)}$$

Proof. Since J_k and μ are multiplicative the two sides of the equality are multiplicative functions of m. Hence the result need only be checked in case m is a prime power p^r. In this case both sides reduce to $\frac{p^k}{p^k - 1}$ and hence the result follows.

Let

(4.3) $$g^{(k)}(m_1,\ldots,m_k,n) = (m_1,\ldots,m_k,n)^k.$$

We then have

Lemma 3.

$$g^{(k)}(m_1,\ldots,m_k,n) = \sum_{d|n} \sigma^{(k)}\left(\frac{n}{d}\right)c(m_1,\ldots,m_k,d),$$

where

(4.4) $$\sigma^{(k)}\left(\frac{n}{d}\right) = \frac{1}{n^k} \sum_{\delta|n} \delta^k c^{(k)}\left(\frac{n}{d}, \frac{n}{d}\right).$$

Proof. It is clear that $g^{(k)}(m_1,\ldots,m_k,n)$ is totally even (mod n). The result now follows from (3.1) and (3.2).

We now consider another arithmetical function $p^{(k)}$, defined by

(4.5) $$p^{(k)}(m_1,\ldots,m_k,n) = \begin{cases} 1 & \text{if } (m_1,\ldots,m_k,n) = 1, \\ 0 & \text{otherwise.} \end{cases}$$

For $k = 1$, this reduces to Kronecker's function (Dickson [1]). We note that this is a multiplicative function. However, the multiplicative properties are not discussed here. It is evident that this is a totally even function (mod n) and has the representation given by (3.1) and (3.2), namely

$$p^{(k)}(m_1,\dots,m_k,n) = \sum_{d|n} \beta_d \, c(m_1,\dots,m_k,d), \tag{4.6}$$

where

$$\beta_d = \frac{1}{n^k} c^{(k)}\left(\frac{n}{d},n\right). \tag{4.7}$$

We will state this as

<u>Lemma</u> 4.

$$p^{(k)}(m_1,\dots,m_k,n) = \frac{1}{n^k}\sum_{d|n} c^{(k)}\left(\frac{n}{d},n\right)c(m_1,\dots,m_k,d).$$

<u>Proof of Theorem 1.</u> Clearly,

$$\begin{aligned} S &= \sum_{(a_1,\dots,a_k,n)=1} (a_1-s_1,\dots,a_k-s_k,n)^k \\ &= \sum_{(a_1,\dots,a_k,n)=1} g^{(k)}(s_1-a_1,\dots,s_k-a_k,n) \\ &= \sum_{\substack{s_i \equiv a_i+b_i \pmod n \\ (i=1,\dots,k)}} p^{(k)}(a_1,\dots,a_k,n)g^{(k)}(b_1,\dots,b_k,n). \end{aligned} \tag{4.8}$$

But by the definition (3.5) of Cauchy composition and (3.7) and Lemmas 3 and 4, we have

$$S = \sum_{d|n} \sigma^{(k)}\left(\frac{n}{d}\right)c^{(k)}\left(\frac{n}{d},n\right)c(s_1,\dots,s_k,d). \tag{4.9}$$

Since $(s_1,\ldots,s_k,n) = 1$, we have (Cohen [5, (4.1)])

$$c(s_1,\ldots,s_k,d) = \mu(d).$$

Therefore (4.8) and (4.9) yield that

$$S = \sum_{d|n} \sigma^{(k)}\left(\frac{n}{d}\right) c^{(k)}\left(\frac{n}{d},n\right)\mu(d).$$

Substituting for $\sigma^{(k)}\left(\frac{n}{d}\right)$ from (4.4) and using Lemma 1 we get that

$$S = \frac{1}{n^k} \sum_{d|n} \left[\sum_{D|\frac{n}{d}} J_k\left(\frac{n}{d}\right) D^k\right] c^{(k)}\left(\frac{n}{d},n\right)\mu(d).$$

Using generalized Hölder's formula (Cohen [5, (2.11)]) for $c^{(k)}$, we obtain from the above that

$$S = \frac{1}{n^k} \sum_{d|n} \left[\sum_{D|\frac{n}{d}} J_k\left(\frac{n}{d}\right) D^k\right] \frac{J_k(n)\,\mu^2(d)}{J_k(d)}$$

$$= \frac{J_k(n)}{n^k} \sum_{D|n} J_k\left(\frac{n}{d}\right) D^k \left(\sum_{d|\frac{n}{D}} \frac{\mu^2(d)}{J_k(d)}\right).$$

This, together with Lemma 2, yields the equality

$$S = \frac{J_k(n)}{n^k} \sum_{D|n} J_k\left(\frac{n}{d}\right) D^k \frac{\left(\frac{n}{d}\right)^k}{J^k\left(\frac{n}{d}\right)},$$

$$J_k(n) \sum_{D|n} 1 = J_k(n)\, d(n).$$

5. Further Lemmata and the Proof of Theorem 2. The following results are needed for the proof of Theorem 2.

Lemma 5. If $d|n$, then

$$\sum_{D|n} D^k c_k\left(\frac{n^k}{d^k}, \frac{n}{D}\right) = \sum_{\delta|\frac{n}{d}} \varphi_k\left(\frac{n}{\delta}\right)\delta^k.$$

Proof. Substituting $\sum_{\delta|\left(\frac{n}{D},\frac{n}{d}\right)} \delta^k \mu\left(\frac{n}{D\delta}\right)$ for $c_k\left(\frac{n^k}{d^k}, \frac{n}{D}\right)$ (see Cohen [3, (1.3)]), we get that the left side is equal to

$$\sum_{D|n} D^k \left[\sum_{\delta|\left(\frac{n}{d},\frac{n}{D}\right)} \delta^k \mu\left(\frac{n}{D\delta}\right)\right]$$

$$= \sum_{\delta|\frac{n}{d}} \delta^k \left(\sum_{D|\frac{n}{\delta}} D^k \mu\left(\frac{n}{D\delta}\right)\right)$$

$$= \sum_{\delta|\frac{n}{d}} \delta^k \varphi_k\left(\frac{n}{\delta}\right)$$

since $\varphi_k(n) = \sum_{d|n} d^k \mu\left(\frac{n}{d}\right)$ (see Cohen [4]).

Lemma 6.

$$\frac{n^k}{\varphi_k(n)} = \sum_{d|n} \frac{\mu^2(d)}{\varphi_k(d)}.$$

Since $\varphi_k = J_k$, this is the same as Lemma 2.

Let us consider another extension of the Kronecker's function, namely

$$(5.1)\qquad p_k(m,n) = \begin{cases} 1 & \text{if } (m,n^k)_k = 1, \\ 0 & \text{otherwise.} \end{cases}$$

The applications of this function to congruence equations are discussed in [9]. We see that $p_k(m,n)$ is a k-even function (mod n^k). Hence using (3.8) and (3.9) we obtain the following

<u>Lemma</u> 7.

$$p_k(m,n) = \frac{1}{n^k} \sum_{d|n} c_k\left(\frac{n^k}{d^k}, n\right) c_k(m,d).$$

Again, let us consider

$$g_k(m,n) = (m,n^k)_k.$$

This is a k-even function (mod n^k). Applying (3.8) and (3.9), we get

$$(5.2)\qquad g_k(m,n) = \sum_{d|n} \alpha_d \; c_k(m,d),$$

where

$$(5.3)\qquad \alpha_d = \frac{1}{n^k} \sum_{\delta|n} g_k(\delta^k,n) c_k\left(\frac{n^k}{d^k}, \frac{n}{\delta}\right)$$

$$(5.3')\qquad = \frac{1}{n^k} \delta^k c_k\left(\frac{n^k}{d^k}, \frac{n}{\delta}\right)$$

$$= \frac{1}{n^k} \sigma_{(k)}\left(\frac{n}{d}\right).$$

where $\sigma_{(k)}\left(\frac{n}{d}\right) = \delta^k c_k\left(\frac{n^k}{d^k}, \frac{n}{\delta}\right)$.

We will state this as

<u>Lemma</u> 8.

$$g_k(m,n) = \frac{1}{n^k} \sum_{d|n} \sigma_{(k)}\left(\frac{n}{d}\right) c_k(m,d).$$

We now present the

<u>Proof of Theorem 2</u>. Consider the Cauchy composite of p_k and g_k, namely

$$\sum_{s \equiv a+b \pmod{n^k}} p_k(a,n)\, g_k(b,n). \tag{5.4}$$

By (3.12) and Lemmas 7 and 8 this is equal to

$$\frac{1}{n^k} \sum_{d|n} \sigma_{(k)}\left(\frac{n}{d}\right) c_k\left(\frac{n^k}{d^k}, n\right) c_k(s,d). \tag{5.5}$$

Since $(s,n^k)_k = 1$, by the definition of p_k and g_k the sum in (5.4) is equal to

$$T = \sum_{(a,n^k)_k = 1} (a-s,n^k)_k. \tag{5.6}$$

In view of the known fact that

$$c_k(s,d) = \mu(d) \quad \text{whenever} \quad (s,d^k)_k = 1,$$

we obtain that

$$T = \frac{1}{n^k} \sum_{d|n} \sigma_{(k)}\left(\frac{n}{d}\right) c_k\left(\frac{n^k}{d^k}, n\right) \mu(d).$$

Substituting for σ_k from (5.3′) and using the generalized Hölder formula (Cohen [3, (1.3)]) we have

$$T = \frac{1}{n^k} \sum_{d|n} \Big[\sum_{D|n} D^k c_k\Big(\frac{n^k}{d^k}, \frac{n}{D}\Big)\Big] \frac{\varphi_k(n)\mu^2(d)}{\varphi_k(d)}$$

$$= \frac{\varphi_k(n)}{n^k} \sum_{d|n} \Big[\sum_{\delta|\frac{n}{d}} \varphi_k\Big(\frac{n}{\delta}\Big)\delta^k\Big] \frac{\mu^2(d)}{\varphi_k(d)}$$

$$= \frac{\varphi_k(n)}{n^k} \sum_{\delta|n} \delta^k \varphi_k\Big(\frac{n}{\delta}\Big) \sum_{d|\frac{n}{\delta}} \frac{\mu^2(d)}{\varphi_k(d)}.$$

In view of Lemma 6, this yields the equality

$$T = \frac{\varphi_k(n)}{n^k} \sum_{\delta|n} \delta^k \varphi_k\Big(\frac{n}{\delta}\Big) \frac{n^k/\delta^k}{\varphi_k\Big(\frac{n}{\delta}\Big)}$$

$$= \varphi_k(n) \sum_{\delta|n} 1 = \varphi_k(n)d(n). \qquad \text{Q.E.D.}$$

In conclusion I wish to refer to [10] for a unitary (Cohen [6]) analogue of Menon's result (1.1) and to Sivaramakrishnan [11] and Subbarao [12] for some related material on (1.1).

REFERENCES

1. L. E. Dickson, History of the Theory of Numbers, Vol. 1, New York, 1934.

2. Eckford Cohen, An extension of Ramanujan's sum, Duke Math. J. 16 (1949), 85-90.

3. Eckford Cohen, An extension of Ramanujan's sum III, Connections with totient functions, Duke Math. J. 23 (1956), 623-630.

4. Eckford Cohen, Some totient functions, Duke Math. J. 23 (1956), 515-522.

5. Eckford Cohen, A class of arithmetical functions in several variables with applications to congruences, Trans. Amer. Math. Soc. 96 (1960), 355-381.

6. Eckford Cohen, Arithmetical functions associated with the unitary divisors of an integer, Math. Z. 74 (1960), 66-80.

7. P. Kesava Menon, On the sum $\sum (a-n)$, $(a,n) = 1$, J. Indian Math. Soc. 29 (1965), 155-163.

8. P. J. McCarthy, The generation of arithmetical identities, J. Reine Angew. Math. 203 (1960), 55-63.

9. K. Nageswara Rao, On a congruence equation and related arithmetical identities, Monatsh. Math. 71 (1967), 24-31.

10. K. Nageswara Rao, Unitary class division of integers (mod n) and related arithmetical identities, J. Indian Math. Soc. 30 (1966), 195-205.

11. R. Sivaramakrishnan, A generalization of an arithmetic function, J. Indian Math. Soc. 33 (1969), 127-132.

12. M. V. Subbarao, Remarks on a paper of P. Kesava Menon, J. Indian Math. Soc. 32 (1968), 317-318.

THE USE OF MEASURE THEORETIC METHODS IN THE STUDY OF ADDITIVE ARITHMETIC FUNCTIONS

C. Ryavec, University of Colorado

By an <u>additive function</u> we shall mean a real-valued, arithmetic function f for which $f(mn) = f(m) + f(n)$ whenever $(m,n) = 1$. Thus, an additive function is determined by its values at the prime powers. Denote the class of additive functions by G.

Much of the investigation into the structure of additive functions has dealt with the distribution of their values. One of the goals of the early research in this direction was to obtain a complete characterization of those functions of G which possess a limiting distribution function, where $f \in G$ has a limiting distribution function $F(x)$ if the finite frequencies

$$n^{-1} \sum_{\substack{m \le n \\ f(m) < x}} 1$$

have a limiting distribution function $F(x)$ (in the probability sense) as $n \to \infty$. This goal was realized in a paper of Erdös and Wintner [2], where they proved that a necessary and sufficient condition that $f \in G$ have a limiting distribution function is that

(1) $$\sum_p \frac{f'(p)}{p} < \infty$$

and

(2) $$\sum_p \frac{(f'(p))^2}{p} < \infty ,$$

where $f'(p) = f(p)$ if $|f(p)| \le 1$ and $f'(p) = 1$ otherwise.

Recently, G. Halász has proved the very important

<u>Theorem</u> (Halász). <u>Let</u> $w(n)$ <u>be a multiplicative function such that</u> $|w(n)| \le 1$. <u>For the limit</u>

$$\lim_{n\to\infty} n^{-1} \sum_{m \le n} w(m)$$

to equal zero, it is necessary and sufficient that either

$$(3) \qquad \sum_p \frac{1 - \mathrm{Re}[w(p)p^{-it}]}{p} = \infty$$

for all real t, *or, if it fails for some* t *(in which case it fails for exactly one* t*) then*

$$\sum_{k=0}^{\infty} \frac{w(2^k)}{2^{(k+it)}} = 0$$

for this t.

(A proof of this theorem may be found in [3].)

The purpose of the present paper is to show, in a particular example, that from Halász' theorem, by an appropriate use of certain results in measure theory, we can deduce new results concerning the distribution of the values of additive functions. We will give, essentially, one half of a theorem due to P. D. T. A. Elliott and myself, which can be found as Lemma 2 of [1]. (For another application of the method see [4].)

We prove the

Theorem. *Suppose that* $f \in G$, *and that there exists a sequence of real numbers* $\alpha_1, \alpha_2, \ldots$ *so that the finite frequencies*

$$n^{-1} \sum_{\substack{m \le n \\ f(m)-\alpha_n < x}} 1 = \nu_n(m: f(m)-\alpha_n < x)$$

possess a limit law as $n \to \infty$. *Then* $f(m)$ *has the form*

$$(4) \qquad f(m) = c \log m + g(m),$$

where $g(m)$ *satisfies condition (2).*

Proof. Let $F(x)$ denote the limit law of the finite frequencies $\nu_n(m: f(m)-\alpha_n < x)$, and let

$$\varphi(t) = \int_{-\infty}^{\infty} e^{itx} dF(x)$$

denote the characteristic function of $F(x)$. Also, put

$$\varphi_n(t) = \int_{-\infty}^{\infty} e^{itx} d\nu_n(m\colon f(m)-\alpha_n < x).$$

Then, if

$$S(n,t) = n^{-1} \sum_{m \le n} e^{itf(m)},$$

we have

$$\varphi_n(t) = S(n,t)e^{-it\alpha_n}. \tag{5}$$

We know that $\lim_{n\to\infty} \varphi_n(t) = \varphi(t)$. Moreover, since $\varphi(t)$ is a characteristic function, $\varphi(0) = 1$ and $\varphi(t)$ is continuous. It follows that $\varphi_n(t)$ fails to converge to zero on some interval about the origin, say on $\Delta = [-\delta,\delta]$, $\delta > 0$. From (5), we see that also $S(n,t)$ fails to converge to zero, as $n \to \infty$, on Δ.

Put

$$G = \{t\colon \lim_{n\to\infty} S(n,t) \neq 0\},$$

where we include in G those real numbers t for which the limit $\lim_{n\to\infty} S(n,t)$ does not exist.

Now for each real number t, $e^{itf(m)}$ is a multiplicative function of absolute value one. Hence, for each $t \in G$ there exists a unique real number $\lambda(t)$ so that the series

$$\sum_p \frac{1 - \mathrm{Re}[e^{itf(m)}\, p^{-i\lambda(t)}]}{p} \tag{6}$$

converges. This follows directly from Halász' theorem.

Let $\|x\|$ denote the distance from x to the nearest integer, and let $f_1(t) \asymp f_2(t)$ signify that there exist positive constants c_1 and c_2 so that $c_1 \le |f_1(t)/f_2(t)| \le c_2$ for all real t. We then have the relations

$$\sum_p \frac{1 - \operatorname{Re}[e^{itf(m)}\, p^{-i\lambda(t)}]}{p}$$

(7)
$$= \sum_p \frac{2\sin^2[\pi(tf(p)/2\pi - \lambda \log p/2\pi)]}{p}$$

$$\asymp \sum_p \frac{\|tf(p)/2\pi - \lambda \log p/2\pi\|^2}{p}.$$

It follows from the convergence of the series in (6) and from the relation (7) that the set G consists precisely of those real numbers t for which there exists a unique real number $\lambda = \lambda(t)$ such that the series in (7) converges.

For each $t \in G$, let $m(t,p)$ denote the nearest integer to

(8)
$$\frac{1}{2\pi}(tf(p) - \lambda \log p);$$

and let $h(t,p)$ be defined by

(9)
$$\frac{1}{2\pi}(tf(p) - \lambda \log p) = m(t,p) + h(t,p),$$

so that $|h(t,p)| \le \frac{1}{2}$. From the convergence of the series in (7) we see that

(10)
$$\sum_p \frac{h^2(t,p)}{p} < \infty.$$

Let t_1 and t_2 be two numbers in G. Set $\lambda(t_1+t_2) \overset{df}{=} \lambda(t_1) + \lambda(t_2)$. Then we see that if $\lambda_i = \lambda(t_i)$, $i = 1, 2$, then

$$\sum_p \frac{\|(t_1+t_2)f(p)/2\pi - (\lambda_1+\lambda_2)\log p/2\pi\|^2}{p}$$

$$\le 2\sum_p \frac{\|t_1 f(p)/2\pi - \lambda_1 \log p/2\pi\|^2}{p}$$

$$+ 2\sum_p \frac{\|t_2 f(p)/2\pi - \lambda_1 \log p/2\pi\|^2}{p},$$

because $\|x+y\|^2 \le 2(\|x\|^2 + \|y\|^2)$, and so $t_1+t_2 \in G$. By the uniqueness of $\lambda(t)$ it follows that $\lambda(t_1+t_2) = \lambda(t_1) + \lambda(t_2)$. A similar argument shows that $\lambda(t_1-t_2) = \lambda(t_1) - \lambda(t_2)$. Therefore, G is an additive group of real numbers. Since $G \supset \Delta$, $G = E_1$.

(Let us note here that we could also arrive at the conclusion $G = E_1$ if we know only that G contains a set of positive measure. This would follow from a theorem of Steinhaus, which says that the set of differences realized in a set of positive measure contains an open interval about the origin.)

The additive property of $\lambda(t)$, coupled with the fact that $\lambda(t)$ is defined on E_1, immediately yields the relation

(11) $$\lambda(rt) = r\lambda(t)$$

for all real numbers t and all rational numbers r. We shall need (11) later.

We shall call a set of primes "thin" if the sum of their reciprocals converges.

Suppose for some nonzero t that $m(t,p) = 0$ for all but a thin set of primes. For such a t, set

$$g(p) = \frac{2\pi}{t}(m(t,p) + h(t,p))$$

and

$$c = \lambda(t)/t.$$

From (8) and (9) we would then obtain

$$f(p) = c \log p + g(p)$$

and

$$\sum_p \frac{(g'(p))^2}{p} < \infty,$$

in which case our theorem is proved.

The remainder of the proof will be devoted to showing that for $t = 1$, $m(1,p) = 0$ for all but a thin set of primes. (In the course of the proof it will emerge that for any t, $m(t,p) = 0$ for all but a thin set of primes. In fact, it will be seen that $\lambda(t)$ satisfies the functional equation $\lambda(t) = t\lambda(1)$.) This will be accomplished in two steps:

First, we will deduce from (9) and (10) that for all real t, there exists a unique $u(t)$ so that the series

$$\sum_p \frac{\|tm(1,p) - u(t)\log p\|^2}{p} \tag{12}$$

converges. Second, we will use (12) to prove that

$$\sum_{\substack{p \\ m(1,p) \neq 0}} \frac{1}{p} < \infty, \tag{13}$$

which will prove the theorem.

To prove (12), rewrite equation (9) with $t = 1$. Thus we have

$$\frac{1}{2\pi}[f(p) - \lambda(1)\log p] = m(1,p) + h(1,p). \tag{14}$$

Multiply equation (14) by t and subtract the product from equation (9), obtaining

$$\begin{aligned} -u(t)\log p &= (2\pi)^{-1}(t\lambda(1) - \lambda(t))\log p \\ &= m(t,p) - tm(1,p) + h(t,p) - th(1,p), \end{aligned} \tag{15}$$

where

$$u(t) = (\lambda(t) - t\lambda(1))/2\pi. \tag{16}$$

Since

$$\sum_p \frac{|h(t,p) - th(1,p)|^2}{p} \le 2\sum_p \frac{h^2(t,p) + t^2h^2(1,p)}{p}$$

$$< \infty,$$

then we see from equation (15) that the series in (12) converges for all real t, where $u(t)$ is given by (16).

We will now show that if $u(t) \equiv 0$ for $0 \le t \le 1$, then equation (13) is true. The proof of the general case will follow from that of the special case, which we state as a

Lemma (W. Schmidt). *Suppose that* $m(1,p)$ *is a sequence of integers such that for every* t *in the unit interval the series*

$$\sum_p \frac{\|tm(1,p)\|^2}{p}$$

converges. Then (13) is true.

Proof. Let $I = [0,1]$, and define $Y(t)$ and $Y_n(t)$ by

$$Y(t) = \sum_p \frac{\|tm(1,p)\|^2}{p}$$

and

$$Y_n(t) = \sum_{p \le n} \frac{\|tm(1,p)\|^2}{p} .$$

For each positive number K, put

$$S(K) = \{t: t \in [0,1]; Y(t) \le K\}.$$

Clearly, $S(K)$ is measurable, and we denote the measure of $S(K)$ by $\mu(K) = \mu(S(K))$. Since

$$\lim_{K\to\infty} \mu(K) = 1,$$

we can choose K so large that $\mu(K) > \frac{3}{4}$, or, what is the same thing, $\mu(I-S(K)) < \frac{1}{4}$.

Then, on the one hand, we have the equation

$$\text{(17)} \qquad \int_0^1 Y_n(t)\,dt = \sum_{p \le n} \int_0^1 \frac{1}{p}\|tm(1,p)\|^2 dt$$

$$= \frac{1}{12} \sum_{\substack{p \le n \\ m(1,p) \ne 0}} \frac{1}{p} ;$$

and on the other hand, we have the estimate

$$\int_0^1 Y_n(t)\,dt = \int_{S(K)} Y_n(t)\,dt + \int_{I-S(K)} Y_n(t)\,dt \tag{18}$$

$$< K + \frac{1}{16} \sum_{\substack{p \le n \\ m(1,p) \ne 0}} \frac{1}{p} .$$

Combining (17) and (18), we see that

$$\frac{1}{12} \sum_{\substack{p \le n \\ m(1,p) \ne 0}} \frac{1}{p} < K + \frac{1}{16} \sum_{\substack{p \le n \\ m(1,p) \ne 0}} \frac{1}{p} ;$$

and it follows that (13) must hold.

To prove that (13) must hold in the general case, we choose an arbitrary constant $k \ge 1$ and a sequence of primes $q = q(p) > p$ satisfying the three conditions

$$q(p_1) = q(p_2) \quad \text{if and only if} \quad p_1 = p_2 \tag{19}$$

$$\lim_{p \to \infty} q/p = k \tag{20}$$

$$\sum_p \frac{\|\log(q/p) - \log k\|^2}{p} < \infty . \tag{21}$$

That a sequence of primes $q(p)$ can be so chosen follows from the fact that there is a prime between x and $x + x\log^{-2} x$ for all sufficiently large x.

We have the inequality

$$\sum_p \frac{\|tm(1,p) - u(t)\log p\|^2}{p}$$

$$\gg \sum_p \left\{ \frac{\|tm(1,p) - u(t)\log p\|^2}{p} + \frac{\|u(t)\log q - tm(1,q)\|^2}{q} \right\}$$

$$\gg \sum_p \frac{\|t(m(1,p) - m(1,q)) - u(t)\log(q/p)\|^2}{p} , \tag{22}$$

where (22) comes from the previous line by adding the p-th term of the first sum and the q-th term of the second sum according to the inequality $\|x+y\|^2 \le 2(\|x\|^2 + \|y\|^2)$, and by noting that $\frac{1}{q} > \frac{1}{2}pk$ for all sufficiently large p. Similarly, by adding the expressions in (22) and (21), letting $v(t) = u(t) \log k$, we see that

$$\sum_p \frac{\|t(m(1,p) - m(1,q)) - v(t)\|^2}{p}$$

$$= \sum_p \frac{\|tm'(p) - v(t)\|^2}{p} \tag{23}$$

converges for all real t, where $m'(p) = m(1,p) - m(1,q)$.

From the convergence of the series in (23) for all t in the unit interval, we shall deduce that $u(t) \equiv 0$. This will prove the theorem, since this case has already been dealt with in the lemma. Henceforth, we assume that $u(t) \not\equiv 0$. We shall arrive at a contradiction.

We first show that for every integer j, the expressions

$$c(j) = \sum_{\substack{p \\ m(p) = j}} \frac{1}{p} \tag{24}$$

are finite. For if not, there exists a j_1 so that the sum defining $c(j_1)$ in (24) diverges. But we see from (23) that

$$\sum_{\substack{p \\ m'(p) = j_1}} \frac{\|tj_1 - v(t)\|^2}{p} < \infty$$

for all $t \in [0,1]$. Therefore,

$$tj_1 - v(t) = n, \tag{25}$$

where $n = n(t,k)$ is an integer. If $j_1 = j_1(k) \equiv 0$, then choose $t \in [0,1]$ so that $u(t) \neq 0$ and choose k so that $v(t) = u(t) \log k$ is not an integer. Then equation (25) cannot hold, so we have a contradiction. (Note that j_1 may depend on k, but that it does not depend on t.) If $j_1 \neq 0$, then choose $t = 1/N$, where N is an integer which does not divide j_1. From (11) we see that $v(1/N) = 0$; and it follows that

$$j_1/N = n \; ,$$

and we again have a contradiction. We conclude that the numbers $c(j)$ defined in (24) are finite.

Now suppose that $c(j) \geq 1$ for infinitely many distinct integers j, say for $j_1, j_2, \ldots$. Then we would have

$$\begin{aligned} \sum_p \frac{\|tm'(p) - v(t)\|^2}{p} &= \sum_{j=-\infty}^{\infty} c(j)\|tj - v(t)\|^2 \\ &\geq \sum_{i=1}^{\infty} c(j_i)\|tj_i - v(t)\|^2 \\ &\geq \sum_{i=1}^{\infty} \|tj_i - v(t)\|^2 \\ &\gg \sum_{i=1}^{\infty} \|t(j_{i+1} - j_i)\|^2 ; \end{aligned}$$

and the last expression cannot converge for all $t \in [0,1]$ by an argument very similar to that given in the proof of the lemma.

Therefore, $c(j) < 1$ for all but finitely many j; and, in fact, we see that $c(j) < \delta$ for all but finitely many integers j for each $\delta > 0$. Hence, we may arrange the numbers $c(j)$ in a non-increasing sequence: $c(j_1) \geq c(j_2) \geq \cdots$.

But

$$\begin{aligned} \sum_p \frac{\|tm'(p) - v(t)\|^2}{p} &= \sum_{i=1}^{\infty} c(j_i)\|tj_i - v(t)\|^2 \\ &\gg \sum_{i=1}^{\infty} c(j_{2i})\|t(j_{2i} - j_{2i-1})\|^2 \end{aligned}$$

and so

(26) $$c(2) + c(4) + c(6) + \cdots < \infty$$

by an argument similar to that used in proving the lemma. Since $c(j_i)$ is a nonincreasing sequence, $(i = 1,2,3,\ldots)$, it follows that

(27) $$c(3) + c(5) + c(7) + \cdots < \infty .$$

Since $c(1)$ is finite, we see from (26) and (27) that

$$\sum_p \frac{1}{p} = \sum_{j=-\infty}^{\infty} c(j) < \infty,$$

which is a contradiction. Therefore $u(t) \equiv 0$ for $t \in [0,1]$, and the theorem is proved.

REFERENCES

1. P. D. T. A. Elliott and C. Ryavec, The distribution of the values of additive arithmetical functions, Acta Mathematica (Sweden) (to appear).

2. P. Erdös and A. Wintner, Additive arithmetical functions and statistical independence, Amer. J. Math. 61 (1939), 713-721.

3. G. Halász, Über die Mittelwerte multiplikativer zahlentheoretischer Funktionen, Acta Math. Acad. Sci. Hungar. 19 (1968), 365-403.

4. C. Ryavec, A characterization of finitely distributed additive functions, J. Number Theory 2 (1970), 393-403.

GENERALIZED ARITHMETIC FUNCTION ALGEBRAS*

David A. Smith, Duke University

1. Introduction. This paper is a progress report on a general method for dealing with the algebraic theory of arithmetic functions with respect to various convolution-type product operations. The method makes use of certain matrix algebras (specifically, incidence algebras and semigroup algebras of semilattices) which are defined in terms of various partial orderings of subsets of the natural numbers. This method and its applications to arithmetic functions have been developed in a recent series of papers by the author [69-73] and, independently, by H. Scheid [60-66]. The points of view are slightly different, and we naturally tend to favor our own, but we have tried to indicate the connections with Scheid's work.

The early work (up to about 1930) on studying arithmetic functions from an algebraic point of view was done by Bell [4-12], Lehmer [48-51], Vaidyanathaswamy [81], and others. The incidence algebra of a lattice was introduced in 1935 by Weisner [83], and was also studied by Ward [82]. More recent, and deeper, results on arithmetic function algebras were obtained by Cashwell and Everett [23] and by Carlitz [19] (and others). At about the same time as the latter, Rota [59] revived the incidence algebra as a tool for the study of combinatorial theory. The author's interest in this area was stimulated by Rota's work. The original intention of our program was to use known facts about arithmetic function algebras to suggest theorems about incidence algebras, which in turn would be applied to combinatorial theory. In fact, the applications in this direction to date have not been very spectacular, but the study has led to a deeper understanding of arithmetic function algebras. In particular, it has shown that many of the important properties of arithmetic functions are based not on the arithmetic of integers but on partial order relations. Thus, for example, it is not necessary to work out "unitary analogues" of Dirichlet-product results because both are based on partial orderings that have in common the property of local distributivity (see Section 11).

Much of the content of this paper has appeared in our earlier papers and Scheid's, but is here reorganized into a more coherent presentation. We also report at length (Section 5) on important recent

*This research was supported in part by NSF grant GP-20092.

work of Doubilet, Rota and Stanley, which has not yet been published. This paper will probably appear in the "Foundations of combinatorial theory" series begun with [59], but in spite of its combinatorial orientation, the sections on reduced incidence algebras contain important insights for arithmetic function algebras.

In Section 2 we list for reference a number of the product operations for arithmetic functions that have been studied. Section 3 contains the definitions of the incidence algebras and semilattice algebras that will be used throughout the paper, and Section 4 lists some inversion formulas which are trivialities from the algebraic point of view.

In Section 5, we identify most of the arithmetic function algebras as reduced incidence algebras and then summarize the results of Doubilet-Rota-Stanley and of Scheid for such algebras.

Sections 6 and 7 introduce the important special functions in incidence algebras and review some results from [72].

The significance of semilattice orderings, with the semilattice operation viewed as a semigroup multiplication, is considered in Section 8. The basic result here is the theorem of Lehmer-von Sterneck that the semigroup algebra of a semilattice is isomorphic to a pointwise algebra of functions, with the isomorphism given by a summation operator. Scheid's generalization of this result is also included. The following section also concerns semilattices, specifically Lindström's theorem on determinants on semilattices and its consequences for arithmetic functions.

In Section 10 we explore the relationship of basic properties of the greatest-integer function to partial orderings. Viewing $[y/x]$, where x and y are positive integers, as the matrix product of characteristic functions of two different partial orderings leads to easy proofs of generalizations of a number of classical identities.

The final three sections review the results of [69, 70, 71] on multiplicative and additive functions, isomorphisms of multiplicative and additive groups in incidence algebras, and generalizations of the identities of Brauer-Rademacher, Hölder, and Landau.

Our list of references is extensive but certainly not exhaustive. A good first approximation to a complete bibliography on this subject would be the union of the reference lists of the items referred to here (minus some irrelevant material).

All unexplained notions having to do with partially ordered sets or lattices can be found in Birkhoff [15].

2. Product operations for arithmetic function algebras. We list for the sake of reference some of the algebra products which have been studied on spaces of arithmetic functions. In each case, f and g are arithmetic functions defined on the non-negative integers $\underline{N}$ or the positive integers $\underline{N}^*$, as appropriate, and the expression given is the value of the product at n.

Cauchy product:

$$\sum_{k=0}^{n} f(k)g(n-k). \tag{2.1}$$

Dirichlet product:

$$\sum_{k|n} f(k)g(n/k). \tag{2.2}$$

Unitary product, introduced by Vaidyanathaswamy [81] and studied extensively by Cohen [28, 29] and others:

$$\sum_{\substack{k|n \\ (k,n/k)=1}} f(k)g(n/k). \tag{2.3}$$

Lucas product, introduced by Carlitz [20, 21]:

$$\sum f(k)g(n-k), \tag{2.4}$$

where the sum is over those k such that a fixed prime $p \nmid \binom{n}{k}$.

<u>Least common multiple</u> (l.c.m.) product, studied principally by Lehmer [49, 50, 51]:

$$(2.5) \qquad \sum_{[k,m]=n} f(k)g(m).$$

<u>Max product</u>, also considered by Lehmer:

$$(2.6) \qquad \sum_{\max(k,m)=n} f(k)g(m).$$

<u>Regular convolution</u>, introduced by Narkiewicz [56]:

$$(2.7) \qquad \sum_{k \in A_n} f(k)g(n/k),$$

where, for each n, A_n is a set of divisors of n subject to conditions which make the product associative, commutative, possess a unit, preserve multiplicativity, and make the appropriate Möbius function (inverse of the constant function 1) take only the values 0 and -1 on prime powers.

<u>Convolution over basic sequences</u>, introduced by Goldsmith [37, 39]:

$$(2.8) \qquad \sum_{\substack{k|n \\ (k,n/k) \in \mathcal{B}}} f(k)g(n/k),$$

where the basic sequence $\mathcal{B}$ is a relation on $\underline{N}^*$ such that:

(i) If $(a,b) \in \mathcal{B}$, then $(b,a) \in \mathcal{B}$,

(ii) $(a,bc) \in \mathcal{B}$ if and only if $(a,b) \in \mathcal{B}$ and $(a,c) \in \mathcal{B}$,

(iii) $(1,b) \in \mathcal{B}$ for all $b \in \underline{N}^*$.

<u>Tau product</u>, introduced by Scheid [62, 65]:

$$(2.9) \qquad \sum_{\tau(k,m)=n} f(k)g(m),$$

where τ is a mapping from a subset of $\underline{N} \times \underline{N}$ into $\underline{N}$ subject to conditions which essentially state that τ is the product operation of a "partially defined" semigroup such that each sum (2.9) has only finitely many terms. (The triple ($\underline{N}$, domain of τ, τ) is an example of what Tainiter [80] calls a generalized semigroup.)

We will not consider in this paper certain other generalized convolution products, such as those involving a kernel function (Gioia [35], Davison [30], Gessley [34], Fredman [33]), nor will we consider the non-associative product studied by Buschman [17].

The motivation for considering (2.7)-(2.9) and other convolution products such as those just mentioned is the desire to be able to deal with several or all of (2.1)-(2.6) simultaneously. One should note that (2.1), (2.2), (2.5), and (2.6) are semigroup algebra products, which is the motivation for (2.9), which in turn includes (2.3) and (2.4) as special cases, even though they are not semigroup algebra products. On the other hand, (2.1)-(2.4) are all incidence algebra products for suitable partial orderings of $\underline{N}$ or $\underline{N}^*$, and the semigroup involved in each of the other two cases is a semilattice. Hence all six cases (as well as (2.7) and (2.8), but not the full generality of (2.9)) can be viewed in terms of partial order relations. This is the point of view that will be emphasized.

3. Algebras of functions of two variables. Let K be a field of characteristic zero (the real numbers $\underline{R}$ or complex numbers $\underline{C}$ will do). Let A be the vector space of functions $f\colon \underline{N} \times \underline{N} \to K$. If we replace such a function f by the array of its values $(f(i,j))$, we may consider f as an infinite matrix over K. Let RF denote the subspace of A of functions which are row-finite in the sense that, for each i, $f(i,j) = 0$ except for finitely many values of j. Similarly, we may define the subspace CF of column-finite functions. If either $f \in$ RF or $g \in$ CF, then we may define the matrix product

$$f * g(x,y) = \sum_{z \in \underline{N}} f(x,z)g(z,y). \tag{3.1}$$

This definition makes both RF and CF into associative K-algebras with identity, and it also makes A into a left unitary RF-module and a right unitary CF-module.

We will use $\leq$ to denote the usual total order on $\underline{N}$. The symbol π will be used to denote partial orderings on subsets of $\underline{N}$ which are coarser than $\leq$. If $P \subseteq \underline{N}$ and π is a partial order on P, then the space $I(\pi,P)$ of <u>incidence functions</u> for π consists of those $f \in A$ such that $f(x,y) = 0$ unless $x \pi y$ in P. No confusion will result from ignoring P and writing I_π for $I(\pi,P)$. This is, of course, a subspace of the space A_P of functions f such that $f(x,y) = 0$ unless $x,y \in P$. I_π consists of upper triangular matrices, and hence is a subalgebra of CF with respect to the matrix product (3.1). I_π is called the <u>incidence algebra</u> of π. Note that the identity element of this algebra is the Kronecker function δ_π on $P \times P$.

<u>Theorem</u> (3.2) (Ward [82]). <u>An element</u> f <u>of</u> I_π <u>is invertible in</u> I_π <u>if and only if</u> $f(x,x) \neq 0$ <u>for all</u> $x \in P$. <u>If this is the case, then for</u> $x,y \in P$,

$$(3.3) \qquad f^{-1}(x,y) = -\Big[\sum_{\substack{x \,\pi\, z \,\pi\, y \\ z \neq y}} f^{-1}(x,z) f(z,y) \Big] \Big/ f(y,y).$$

We denote ζ_π the characteristic function of the relation π: $\zeta_\pi(x,y) = 1$ if $x \pi y$ in P, 0 otherwise. Then ζ_π is invertible in I_π, and its inverse is called the <u>Möbius function</u> of π denoted by μ_π. From (3.3), we have

$$(3.4) \qquad \mu_\pi(x,y) = - \sum_{\substack{x \,\pi\, z \,\pi\, y \\ z \neq y}} \mu_\pi(x,z).$$

(Both (3.3) and (3.4) have obvious dual statements which we omit.)

We have defined incidence algebras in such a way that, for any partial order coarser than $\leq$, I_π is a subalgebra of $I(\leq,\underline{N})$. It is useful to observe a more general relationship for the incidence algebras of related partial orders. The proof is trivial.

<u>Theorem</u> (3.5). <u>If</u> $P_1 \subseteq P_2 \subseteq \underline{N}$, <u>and</u> π_1 <u>and</u> π_2 <u>are partial orders on</u> P_1 <u>and</u> P_2, <u>respectively, such that</u> $x \pi_1 y$ <u>implies</u> $x \pi_2 y$ <u>for all</u> $x,y \in P_1$, <u>then</u> $I(\pi_1,P_1)$ <u>is a subalgebra of</u> $I(\pi_2,P_2)$.

Important special cases of (3.5) include the case in which $P_1 = P_2$ and that in which the sets are different, but π_1 is the restriction to P_1 of π_2.

There are a number of useful ways to consider functions of one variable on a subset P as a subspace of A_P. Some of these will be considered in Sections 5 and 8, but we mention here the space B_P of functions of the first variable only: $f(x,y) = f(x,x)$ for all $x,y \in P$. Whenever $f \in B_P$, we write $f(x,y) = f(x)$. The space of functions of the second variable only may be considered as the set B_P^t of transposes of elements of B_P.

If π is a partial ordering of a subset P which is a lower semilattice ordering (i.e., infima exist), and if each $x \in P$ can be written as $x = u \wedge v$ in only finitely many ways, then we may define the <u>wedge product</u>

$$(3.6) \qquad f \wedge g(x,y) = \sum_{u \wedge v = x} f(u,y)g(v,y)$$

for $f,g \in A_P$. The restriction of this product to B_P is just the semigroup algebra product of the semilattice $(P,\wedge)$.

Similarly, if π is an upper semilattice ordering coarser than $\leq$ (which implies the necessary finiteness condition), then we define

$$(3.7) \qquad f \vee g(x,y) = \sum_{u \vee v = y} f(x,u)g(x,v),$$

for $f,g \in A$. The restriction to B_P^t is the semigroup algebra product of $(P, \vee)$.

The <u>pointwise product</u> of functions in A will be denoted by juxtaposition:

$$(3.8) \qquad fg(x,y) = f(x,y)g(x,y).$$

A is, of course, a commutative K-algebra with respect to this product, and the various subspaces RF, CF, A_P, I_π, B_P are all subalgebras. Indeed, all of these except B_P are ideals.

An invertible incidence function f in I_π is called <u>completely multiplicative</u> (or <u>completely factorable</u>) if

(3.9) $$f(x,y)f(y,z) = f(x,z) \quad \text{whenever} \quad x \;\pi\; y \;\pi\; z.$$

It is immediately evident that if f is completely multiplicative and g and h are abitrary in I_π, then

(3.10) $$f(g * h) = (fg) * (fh).$$

Additional references: Apostol [3], Rota [59], Smith [69 70, 73], Stanley [75], Weisner [83].

4. <u>Inversion formulas</u>. Let π be a partial order on $P \subseteq \underline{N}$, refined by $\leq$. We write I, ζ, μ for $I_\pi, \zeta_\pi, \mu_\pi$, since π is fixed. The Möbius Inversion Formula for the partial order π is really just the statement that μ is the inverse of ζ in I: For f, F in A_P, we have

(4.1) $$F = f * \zeta \text{ if and only if } f = F * \mu.$$

Suppose $P_1 \subseteq P_2$, π is a partial order on P_2, and P_1 is ordered by restriction of π. Write δ_i, ζ_i, μ_i for the appropriate functions on (P_i, π). We define

(4.2) $$\mu' = \delta_1 * \mu_2, \quad \overline{\mu} = \zeta_1 * \mu_2.$$

These functions may be used for partial inversion formulas, as follows: If $f \in I(\pi, P_2)$ and $F = \zeta_2 * f$, then

(4.3) $$\mu' * F(x,y) = f(x,y) \text{ if } x \in P_1, \ 0 \text{ otherwise;}$$

(4.4) $$\overline{\mu} * F(x,y) = \sum_{\substack{x \,\pi\, z \,\pi\, y \\ z \in P_1}} f(z,y) \text{ if } x \in P_1, \ 0 \text{ otherwise.}$$

All of the above formulas have dual formulations which we omit.

Some authors (e.g., Hsu [43]) have studied "higher order" inversion formulas. Hsu's functions μ_m are obtained by inverting chain-counting functions, which are just powers of ζ under $*$ multiplication. Thus higher order inversion is obtained by iterating (4.1):

$$(4.5) \qquad F = f * \zeta^m \text{ if and only if } f = F * \mu_m.$$

Additional references for inversion formulas: Cashwell-Everett [23], Cohen [27], Gupta [40], Hardy-Wright [41], Smith [69, 72].

5. Reduced incidence algebras. We commented in Section 2 that most of the familiar products of arithmetic functions are incidence algebra products. We now elaborate on that claim to illustrate the concept of reduced incidence algebra. (The other product operations, such as (2.5) and (2.6), which are semigroup algebra products, will be considered in Section 8.)

Example (5.1). Let $T = I(\leq, \underline{N})$ be the largest incidence algebra on $\underline{N}$, namely the algebra of upper triangular matrices. Define an equivalence relation ρ on the set of intervals in $\underline{N}$: $[x,y]\ \rho\ [z,w]$ if $y-x = w-z$. Define I_C to be the subspace of T of functions f which are constant on ρ-classes. The classes may be identified with the elements of $\underline{N}$ in an obvious way: the class of $[x,y]$ is its length $y-x$. For functions $f \in I_C$ we may write $f(x,y) = f(y-x)$ and identify f with a function of one variable on N. With this identification, I_C becomes the algebra of arithmetic functions with the Cauchy product (2.1), or equivalently, the algebra of sequences with convolution product, or the algebra of formal power series.

Example (5.2). Let $\underline{N}^*$ be partially ordered by divisibility. Then an interval $[x,y]$ is the set of z such that $x|z$ and $z|y$. Define an equivalence relation ρ on the set of intervals by $[x,y]\ \rho\ [z,w]$ if $y/x = w/z$. Define I_D to be the subset of $I(|,\underline{N}^*)$ of functions which are constant on ρ-classes. If we identify the class of $[x,y]$ with the integer y/x and $f(x,y)$ with $f(y/x)$, I_D becomes the algebra of arithmetic functions with the Dirichlet product (2.2), or by a standard identification, the algebra of formal Dirichlet series.

<u>Example</u> (5.3). Let $\underline{N}^*$ be ordered by unitary divisibility:

$$x \pi y \text{ if and only if } x|y \text{ and } (x,y/x) = 1.$$

If $x \pi y$, the interval $[x,y]$ consists of those z which are multiples of x, divisors of y, and $(x,z/x) = (z,y/z) = 1$. If $y/x = \prod_1^n p_i^{\alpha_i}$, where the p_i are distinct primes, it is easy to see that the z's in $[x,y]$ are precisely the products of $p_i^{\alpha_i}$ over subsets of the index set $\{1, 2, \ldots, n\}$. Thus each interval $[x,y]$ is order-isomorphic to a finite Boolean algebra.

If we define ρ exactly as in Example (5.2), but for unitary intervals, and let I_u be the set of $f \in I_\pi$ which are constant on ρ-classes, then it is easy to see that I_u is isomorphic to the algebra of arithmetic functions with the unitary product (2.3).

<u>Example</u> (5.4). Let p be a fixed prime, and let $\underline{N}$ be partially ordered by the relation

$$x \pi y \text{ if and only if } p \nmid \binom{y}{x}.$$

We note for future reference that $(\underline{N},\pi)$ is order-isomorphic to the subset of $(\underline{N}^*,|)$ consisting of integers with no p-th power factor. If we write an arbitrary $n \in \underline{N}$ as $\sum a_k p^k$, $0 \le a_k < p$, the isomorphism is given by

$$\nu(n) = \prod p_k^{a_k}, \tag{5.5}$$

where $p_0, p_1, p_2, \ldots$ are the primes in order.

If we define ρ as in Example (5.1) and I_p as the subspace of functions constant on ρ-classes, then I_p is isomorphic to the algebra of arithmetic functions with the Lucas product (2.4) corresponding to p.

The isomorphism ν of $(\underline{N},\pi)$ with a subset of $(\underline{N}^*,|)$ induces an isomorphism of I_p with a subalgebra of I_D, which has been noted by Carlitz [21].

<u>Example</u> (5.6). Convolution over basic sequences (2.8) includes Dirichlet and unitary convolutions as special cases. Exactly as in Examples (5.2) and (5.3), one may partially order $\underline{N}^*$ by a basic sequence $\mathcal{B}$:

$$x \pi y \text{ if and only if } x|y \text{ and } (x, y/x) \in \mathcal{B}.$$

Then functions which are constant on intervals with constant quotients form a subalgebra $I_{\mathcal{B}}$ of I_{π} which is just the convolution algebra with product (2.8). (This has been noted by Goldsmith in [39].)

The preceding examples all have in common an equivalence relation ρ on intervals of a partially ordered set which has the property that the set of functions constant on ρ-classes is closed under the matrix product (3.1). Such an equivalence relation is called an <u>order-compatible</u> relation. An example of an order-compatible relation on an arbitrary partial ordering is the relation of isomorphism of intervals as partially ordered sets (Smith [69], Scheid [63]). The subalgebra I_{ρ} of the incidence algebra consisting of functions constant on ρ-classes is called a <u>reduced incidence algebra</u>. The terminology of this paragraph and most of the known results on reduced incidence algebras are due to Doubilet, Rota, and Stanley [32], and their results will be sketched in this section, along with those of Scheid [63-65].

The examples above raise some obvious questions about reduced incidence algebras, in light of known properties of arithmetic function algebras. For example:

(a) How does one recognize an order-compatible relation? Can they be characterized internally with respect to the partial ordering, i.e., without reference to incidence functions?

(b) Is isomorphism necessarily coarser than every other order-compatible relation on the intervals of a given partial ordering?

(c) When is I_{ρ} commutative? (All the arithmetic function examples are.)

(d) When is I_{ρ} an integral domain? (I_C and I_D are, but I_u and I_P are not.)

With reference to question (a), we have the following sufficient condition.

<u>Theorem</u> (5.7) (Smith [70]). <u>Let</u> (P,π) <u>be a partially ordered set and</u> ρ <u>an equivalence relation on intervals such that, whenever</u> $[x,y]\ \rho\ [z,w]$, <u>there exists a bijection</u> $\varphi: [x,y] \to [z,w]$ <u>such that</u> $[u,v]\ \rho\ [\varphi(u),\varphi(v)]$ <u>for all</u> u, v <u>in</u> $[x,y]$. <u>Then</u> ρ <u>is order-compatible</u>.

The sufficient condition of (5.7) was our definition of compatibility in [70] in an attempt to answer question (a) that was doomed to failure. The bijection φ in (5.7) is necessarily an order-isomorphism, and it happens that the answer to question (b) is "no", as we shall see. Hence the condition of (5.7) cannot be necessary. Doubilet, Rota, and Stanley [32] finessed question (a) by giving the definition of compatibility in terms of closure of products of the appropriate functions. The second part of question (a) remains unanswered.

The results numbered (5.8)-(5.20) are all taken from Doubilet-Rota-Stanley [32], where proofs may be found.

<u>Theorem</u> (5.8). <u>Given</u> (P,π), <u>the family of order-compatible relations on intervals of</u> P, <u>partially ordered by refinement of partitions, forms a complete lattice, hence has a largest (coarsest) element</u>.

The following Hasse diagram for a partial ordering on a set P of 16 elements shows that the maximal order-compatible relation may be coarser than the isomorphism relation, answering question (b) in the negative:

The maximal relation identifies the two components (which are obviously not order-isomorphic), and otherwise agrees with the isomorphism relation.

We now fix an order-compatible relation ρ, and label the ρ-classes of intervals $\alpha, \beta, \gamma, \ldots$. For arbitrary classes α, β, γ, we define

$$\begin{bmatrix} \alpha \\ \beta, \gamma \end{bmatrix}$$

to be the number of distinct elements z in an interval [x,y] of class α such that [x,z] is of class β and [z,y] is of class γ. It is easy to see that this is independent of the choice of [x,y] in the class α. If we now identify a function $f \in I_\rho$ with a function defined on ρ-classes (i.e., $f(\alpha) = f(x,y)$ where [x,y] is any interval of class α), then it is easy to verify that the product in I_ρ is given by

$$(5.9) \qquad f * g(\alpha) = \sum_{\beta,\gamma} \begin{bmatrix} \alpha \\ \beta, \gamma \end{bmatrix} f(\beta) g(\gamma).$$

If ρ_1 and ρ_2 are order-compatible relations, and ρ_1 refines ρ_2, then I_{ρ_2} is clearly a subalgebra of I_{ρ_1}. In particular, if we write $\overline{I}$ (or $\overline{I}_\pi$) for the reduced incidence algebra of the maximal order-compatible relation, $\overline{I}$ is a subalgebra of every reduced incidence algebra. $\overline{I}$ is called the <u>maximally</u> <u>reduced</u> incidence algebra. We denote by I^* the reduced incidence algebra corresponding to the isomorphism relation and call it the <u>standard</u> reduced incidence algebra. I^* is a subalgebra of every I_ρ for ρ satisfying the condition of Theorem (5.7).

<u>Theorem</u> (5.10). <u>If</u> ρ_1 <u>is a finer order-compatible relation than</u> ρ_2 <u>and</u> f <u>is an element of</u> I_{ρ_2} <u>which is invertible in</u> I_{ρ_1}, <u>then</u> f <u>is invertible in</u> I_{ρ_2}.

Since $\zeta_\pi \in \overline{I}_\pi$ and is invertible in I_π (a trivially reduced incidence algebra), it follows that $\mu_\pi \in \overline{I}_\pi$, hence is in every reduced incidence algebra. Theorem (5.10) also implies Theorem 8 of [70].

Theorem (5.11). *If* (P,π) *is a finite partially ordered set and* S *is a subalgebra of* I_π *containing the identity* δ_π, *then* S *is a reduced incidence algebra if and only if* $\zeta_\pi \in S$ *and* S *is closed under pointwise multiplication*.

If the base field K is a topological field, then I_π can be given the topology of pointwise convergence. The characterization of reduced incidence algebras when P is infinite is similar to (5.11) but requires in addition that S be a topologically closed subalgebra.

In view of the fact that $\bar{I}_\pi$ need not equal I^*_π, it is of interest to consider in more detail which intervals in (P,π) are equivalent in the maximal order-compatible relation. We call such intervals *residually isomorphic*. (This is a deliberate oversimplification of some important details in [32], which is possible only because proofs are being omitted.)

Theorem (5.12). *Residually isomorphic intervals have the following properties in common:*

(i) *number of maximal chains of a given length*;

(ii) *number of elements of given minimal length from the bottom (or top) element (for example, number of atoms or dual atoms)*;

(iii) *number of elements*.

Question (c), concerning commutativity of I_ρ, has an obvious, but not very satisfactory answer: It follows from the expression (5.9) for the product that I_ρ is commutative if and only if

$$\begin{bmatrix} \alpha \\ \beta,\gamma \end{bmatrix} = \begin{bmatrix} \alpha \\ \gamma,\beta \end{bmatrix} \tag{5.13}$$

for all ρ-classes α, β, γ. For $\bar{I}$ we can give a better answer. An interval is called *residually self-dual* if it is residually isomorphic to the finite partially ordered set obtained by reversing the order relation (the *dual* interval).

Theorem (5.14). $\bar{I}_\pi$ *is commutative if and only if every interval of* (P,π) *is residually self-dual*.

An interval is called <u>self-dual</u> if it is order-isomorphic to its dual. The author noted in [69] that self-duality of every interval is sufficient for I^* to be commutative, and raised the question of necessity. Scheid [63] showed that this condition is not necessary, and gave the necessary and sufficient condition (5.13) for the isomorphism case.

The following example is the Hasse diagram of a partially ordered set which is residually self-dual (as is every subinterval) but not self-dual. It is easy to see that $\bar{I} = I^*$ in this case, so I^* is commutative.

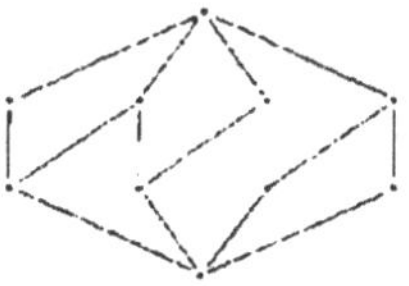

In view of the arithmetic function examples, it is of interest to ask which reduced incidence algebras are isomorphic to the algebra of formal power series or to the algebra of formal Dirichlet series. The answers to these, of course, have a bearing on questions (c) and (d) above.

Let I_ρ be a reduced incidence algebra on (P,π) such that the ρ-classes are in one-to-one correspondence with a subset of $\underline{N}^*$, the class of $[x,y]$ being denoted $\rho(x,y)$, so that ρ becomes an element of I_ρ. If ρ is completely multiplicative (see (3.9)), we say I_ρ is an <u>algebra of Dirichlet type</u>. An algebra I_ρ of Dirichlet type is <u>full</u> if, whenever n is a ρ-class and $k|n$, $\rho(x,y) = n$ implies that there exists $z \in [x,y]$ such that $\rho(x,z) = k$, $\rho(z,y) = n/k$. I_ρ is said to be of <u>binomial type</u> if all the classes $\rho(x,y)$ are powers of a fixed prime p.

Let I_ρ be a full <u>commutative</u> algebra of Dirichlet type. Suppose $n = \rho(x,y)$ and $n = p_1 p_2 \cdots p_m$ is any ordered factorization of n into primes. One can show that the number of chains

$$x = x_0 < x_1 < \cdots < x_m = y$$

such that $\rho(x_{i-1},x_i) = p_i$, for each i, depends only on n; call this number $b(n)$.

Theorem (5.15). *If* I_ρ *is a full commutative algebra of Dirichlet type with* ρ-*classes* $1 = n_1, n_2, \ldots,$ *then the map*

$$f \to \sum_k \frac{f(n_k)}{b(n_k)n_k^s}$$

is an isomorphism of I_ρ *into the algebra of formal Dirichlet series with multiplication modulo all terms involving* n^{-s} *where* n *is not a* ρ-*class*.

For the case of binomial type algebras, much more can be said. If the classes are all labeled p^α for some α, it is convenient to replace p^α by α, so that ρ becomes completely additive. I_ρ is then full if, whenever $\rho(x,y) = n$ and $0 \le k \le n$, there exists z such that $\rho(x,z) = k$, $\rho(z,y) = n-k$.

Theorem (5.16). *Every algebra of full binomial type is commutative*.

Let I_ρ be of full binomial type, let $a(n)$ be the number of atoms in an interval $[x,y]$ of class n, and let $b(n) = a(1)a(2)\cdots a(n)$. Let M denote the largest class $\rho(x,y)$, if there is one, or $M = \infty$ if not.

Theorem (5.17) *If* I_ρ *is of full binomial type, then the map*

$$f \to \sum_{n=0}^{M} \frac{f(n)}{b(n)} t^n + (t^{M+1}) \tag{5.18}$$

is an isomorphism of I_ρ *with* $K\langle t\rangle/(t^{M+1})$, *where* $K\langle t\rangle$ *is the formal power series algebra*.

The next theorem completely characterizes algebras I_ρ which have an isomorphism of the form (5.18). In particular, ρ must be the maximal order-compatible relation on a partial ordering of a very restricted sort. If an isomorphism of the form (5.18) exists, it can always be normalized so that $b(1) = 1$ by replacing t by $t' = b(1)^{-1}t$.

Theorem (5.19). *Suppose* I_ρ *is a reduced incidence algebra on* P *with* ρ*-classes labelled* $0, 1, 2, \ldots, M$ *(finite or infinite) and* b *is a function such that* $b(1) = 1$ *and (5.18) gives an isomorphism between* I_ρ *and* $K<t>/(t^{M+1})$. *Then*

(i) P *satisfies the Jordan-Dedekind chain condition;*

(ii) *all intervals in* P *of length* n *have the same number of maximal chains;*

(iii) *if* $[x,y]$ *has length* n, *then* $\rho(x,y) = n$;

(iv) $b(n)$ *is the number of maximal chains in an interval of length* n, *and* M *is the length of* P;

(v) $I_\rho = \bar{I}(P)$.

Conversely, if P *satisfies (i) and (ii), then* $\bar{I}(P)$ *is an algebra of full binomial type given by (iii) and (iv).*

Theorem (5.20). *Let* P *be a lattice satisfying (i) and (ii) of (5.19), with the additional property that the join of any two atoms has height 2. Then* P *is isomorphic to one of the following:*

(a) *a chain;*

(b) *the lattice of finite subsets of a set;*

(c) *the lattice of finite subspaces of a vector space.*

The Cauchy and Dirichlet algebras of arithmetic functions (Examples (5.1) and (5.2)) are, of course, the prototypes of full algebras of binomial and of Dirichlet types, respectively. The maximally reduced algebra for $(\underline{N}^*, |)$, which is the same as the standard reduction, is also of Dirichlet type for a suitable labelling of the isomorphism classes, but is not full for any such labelling. The maximal reduction for the unitary ordering (Example (5.3)) is of full binomial type for a suitable labelling of classes. The Lucas algebra (5.4) for a fixed prime p is of full Dirichlet type with respect to the assignment $\rho(x,y) = \nu(y-x)$ given by (5.5). As with the divisibility ordering, the maximal reduction is Dirichlet, but not full. For the case of convolution over basic sequences (5.6), the algebra $I_{\mathcal{B}}$ is of full Dirichlet type if the assignment is $\rho(x,y) = y/x$. However, the nature of the maximal reduction depends on the choice of $\mathcal{B}$, as the divisibility and unitary cases show.

The question of absence of zero-divisors in the standard reduced incidence algebra has been considered by Scheid [63, 64, 65] in the case in which (P, π) is a direct product of trees. In this case, $I^* = \overline{I}$. The following results are taken from [64].

Theorem (5.21). *If* P *is the direct product of finitely many trees* T_i, *then* $\overline{I}$ *has no (non-zero) zero-divisors if and only if each* T_i *has infinite length.*

Theorem (5.22). *If* P *is the countable direct product of trees* T_i *then* $\overline{I}$ *has no zero-divisors if and only if, for each* T_j *of finite length, each chain of* T_j *has an isomorphic copy in infinitely many* T_i.

The condition in (5.22) is equivalent to at least one of the following being infinite: the supremum m of lengths of T_i's; the number of T_i of length m; or the number of T_i of infinite length.

In the same paper Scheid gives the following results on prime factorization in the algebra $\overline{I}_\pi$, where (P, π) is a direct product of chains.

Theorem (5.23). $\overline{I}_\pi$ *has only one prime (up to associates) in each of the following cases:* $(P, \pi) \cong (\underline{N}, \le)$ *or* $(P, \pi) \cong$ *the set of finite subsets of* $\underline{N}$, *ordered by inclusion.*

Theorem (5.24). *If* $(P, \pi) \cong (\underline{N}, \le) \times \cdots \times (\underline{N}, \le)$, r factors, with $r \ge 2$, *then* $\overline{I}_\pi$ *has more than* r *primes (up to associates), and every element has a prime factorization.*

Theorem (5.25). *For the case of* $(\underline{N}^*, |)$ *(a direct product of infinitely many infinite chains),* $\overline{I}$ *has infinitely many primes, and every element has a prime factorization.*

Scheid's method in proving these latter results resembles that of Cashwell and Everett [23] for the Dirichlet algebra I_D. However, the method of [23] for proving that I_D is a unique factorization domain does not appear to work for $\overline{I}$. Scheid poses as open problems for $\overline{I}$, in the cases of (5.24) and (5.25), the determination of all the primes, and settling uniqueness of prime factorization.

6. Special functions. Every incidence algebra I_π contains some functions which are more interesting than others, of which we have already mentioned δ, ζ, μ. We also noted in §4 that $\zeta^k(x,y)$ is the number of chains of length k in $[x,y]$. In particular, we write $\tau = \zeta^2$: $\tau(x,y) = |[x,y]|$. Some others introduced by Ward [82]:

$$\varkappa(x,y) = \begin{cases} 1, & \text{if } x = y \text{ or } y \text{ covers } x, \\ 0, & \text{otherwise.} \end{cases} \tag{6.1}$$

(6.2) $\lambda(x,y)$ = number of dual atoms in $[x,y]$.

(6.3) $\lambda'(x,y)$ = number of atoms in $[x,y]$.

(6.4) $d(x,y)$ = length of the longest chain in $[x,y]$.

The last is primarily of interest when P satisfies the Jordan-Dedekind chain condition, in which case $d(x,y) = d(y) - d(x)$, where the latter d is the dimension function in P (see Birkhoff [15, pp. 11, 67]). One verifies easily that

$$\lambda = \zeta * (\varkappa - \delta), \qquad \lambda' = (\varkappa - \delta) * \zeta. \tag{6.5}$$

In order to develop totient functions and others, it is necessary to have an appropriate notion of powers of an element of P. Since we are considering only subsets of $\underline{N}$, which may in turn be considered as a subset of K, there is an appropriate notion at hand. However, this is not sufficiently general to account, for example, for the Lucas analogue of the Euler function. Thus, we assume as given a function $\nu: P \to K^*$, which may be quite arbitrary. In the cases of Examples (5.2) and (5.3), the identity function on $\underline{N}^*$ is quite appropriate. In (5.1), one might consider $\nu(n) = 2^n$, and in (5.4) the map given by formula (5.5). Corresponding to ν we have an incidence function of the same name:

$$\nu(x,y) = \nu(y)/\nu(x), \tag{6.6}$$

which is obviously completely multiplicative, in the sense of (3.9). We denote the pointwise powers of ν by subscripts:

$$\nu_k(x,y) = \nu(y)^k/\nu(x)^k. \tag{6.7}$$

The functions ν_k are invertible in I_π, and

(6.8) $$\nu_k^{-1} = \nu_k \mu.$$

(This follows from (3.10) by setting $f = \nu_k$, $g = \zeta$, $h = \mu$.) If the field K is one in which real (non-integral) powers exist, (6.7) can be used as the basis for more general functions such as those defined by Jager [45], but we will not bother with this.

The generalized sum-of-divisors and totient functions are defined by

(6.9) $$\sigma_k = \nu_k * \zeta,$$

(6.10) $$\varphi_{j,k} = \nu_j * \nu_k^{-1} = \nu_j * \nu_k \mu.$$

The latter include the Jordan totient functions (in particular, the Euler φ-function), as well as their unitary and Lucas analogues, and many others. Familiar properties of these functions are immediate consequences of the definitions and computations of their inverses:

(6.11) $$\sigma_k^{-1} = \mu * \nu_k \mu,$$

(6.12) $$\varphi_{j,k}^{-1} = \varphi_{k,j}.$$

(See [69].)

The generalized Euler function is $\varphi_{1,0}$, which we abbreviate φ. The Jordan functions are $\varphi_{k,0} = \varphi_k$. Thus

(6.13) $$\varphi_k = \nu_k * \mu,$$

and in particular,

(6.14) $$\varphi = \nu * \mu.$$

The functions φ_k are not as general as those introduced recently by Stevens [77], but his functions do have the form $\nu_k * \Omega\mu$, where Ω is a completely multiplicative function.

Easy computations show that

(6.15) $$\varphi_k * \varphi_j^{-1} = \varphi_{k,j} = \varphi_{k-j}\nu_j ,$$

and

(6.16) $$\sigma_k * \sigma_j^{-1} = \varphi_{k-j}\nu_j .$$

These identities are attributed to Gegenbauer and Negoes, respectively, in the Dirichlet case, and include the unitary case (Jager [45]).

Other generalized totient functions depend on consideration of closure operators on partially ordered sets, and will be considered in the next section.

7. Closure operators. As before, let (P,π) be a partially ordered set, where $P \subseteq N$ and $\le$ refines π. Let $x \to \bar{x}$ be a closure operator on P, and let Q be the subset of closed elements, ordered by the restriction of π. We will write δ, μ, ζ, etc., for the appropriate functions in $I(\pi,P)$, and δ_Q, μ_Q, ζ_Q, for the corresponding functions in $I(\pi,Q)$. The following results are taken from Smith [72].

Let $\bar{\delta} \in I(\pi,P)$ be defined by $\bar{\delta}(x,y) = \delta(\bar{x},y)$. Then right multiplication by $\bar{\delta}$ has the following effect:

(7.1) $$f * \bar{\delta}(x,y) = \sum_{\bar{z}=y} f(x,z).$$

Theorem (7.2) $$\bar{\delta} = \zeta * \mu_Q .$$

This easily proved identity has a number of immediate and interesting consequcnces:

(7.3) $$\mu * \bar{\delta} = \mu_Q \qquad \text{(Rota [59]).}$$

(7.4) $$\zeta * \bar{\delta} = \tau * \mu_Q .$$

The left-hand member of (7.4) counts the number of z in $[x,y]$ such that $\bar{z} = y$. The right-hand member uses the counting function τ

for whole intervals in P.

As a subset of P, Q determines functions μ' and $\overline{\mu}$ defined by (4.2):

$$\mu' = \delta_Q * \mu, \quad \overline{\mu} = \zeta_Q * \mu.$$

We have:

(7.5) $$\mu' * \overline{\delta} = \mu_Q.$$

(7.6) $$\overline{\mu} * \overline{\delta} = \delta_Q.$$

Corresponding to a map $\nu: P \to K^*$, as in Section 6, both P and Q have σ-functions and φ-functions defined by (6.9) and (6.13).

(7.7) $$\varphi_k * \overline{\delta} = \varphi_{k,Q}.$$

(7.8) $$\sigma_k * \overline{\delta} = \nu_k * \tau * \mu_Q.$$

If ν_k and τ_Q commute,

(7.9) $$\tau_Q * \varphi_k * \overline{\delta} = \sigma_{k,Q}.$$

For the chain-counting functions, we have:

(7.10) $$\zeta^k * \overline{\delta} = \zeta^{k+1} * \mu_Q, \quad \text{for all } k.$$

(7.11) $$\zeta_Q^k * \mu * \overline{\delta} = \zeta_Q^{k-1}, \quad \text{for all } k.$$

An interesting example of the use of these identities is obtained by letting $P = \underline{N}^*$, ordered by divisibility, and $\overline{x}$ = the smallest r-th power multiple of x, where r is a fixed positive integer. Then Q is the set of r-th powers. Since $(\underline{N}^*, |) \cong (Q, |)$ under the map $x \to x^r$, we have $\mu_Q(x^r, y^r) = \mu(x,y) = \mu(y/x)$. We will restrict our attention to the reduced incidence algebra I_D, i.e., the ordinary algebra of arithmetic functions with Dirichlet product.

The number of elements $k \in \underline{N}^*$ such that $\overline{k} = m^r$ is clearly $r^{\lambda(m)}$, where $\lambda(m)$ is the number of distinct prime divisors of m. By (7.4), we have:

$$(7.12) \qquad {}_r\lambda(m) = \sum_{d|m} \tau(d^r)\,\mu(m/d),$$

where $\tau(k)$ is the number of divisors of k. This is equivalent to:

$$(7.13) \qquad \tau(m^r) = \sum_{d|m} {}_r\lambda(d).$$

Note that no use of factorability is made in obtaining these identities.

Taking ν to be the identity function, φ_k is the Jordan totient function and $\sigma_k(m)$ is the sum of the k-th powers of divisors of m. The function $\varphi_{1,Q}$ is Klee's generalization of the Euler function [46], so $\varphi_{k,Q}$ may be viewed as a Klee-Jordan function. ζ^k is the function denoted τ_k by Beumer [14] and ζ_Q^k is the generalization $\tau_{k,r}$ introduced by Sivaramakrishnan [68]. We also have $\bar{\mu}(1,m) = v_r(m)$, where v_r is the generalized Möbius function of Gupta [40]. The partial inversion formula (4.4) in this case is Theorem 3 of [40]. The reader is invited to write out the identities for these functions that may be obtained from (7.6)-(7.11).

In a recent paper, Scheid and Sivaramakrishnan [67] have given some identities involving many of these same functions, as well as their unitary and l.c.m. analogues. This work does not proceed from closure operator considerations, but rather from the fact that $x \to x^r$ is an isomorphism from $(\underline{N}^*, |)$ to a subset of itself (proper unless $r = 1$). Given (P,π) and an isomorphism $\Phi: P \to Q \subseteq P$, we have an induced isomorphism $\Omega: I(P) \to I(Q) \subseteq I(P)$ given by $\Omega f(\Phi(x),\Phi(y)) = f(x,y)$. Of course, $\Omega\zeta = \zeta_Q$, so $\Omega\zeta^k = \zeta_Q^k$ and $\Omega\mu = \mu_Q$. In the case of the r-th power map, the source of unitary analogues is clear, since the incidence algebra for the unitary ordering (5.3) is a subalgebra of that for the divisibility ordering, by Theorem (3.5). The l.c.m. analogues proceed from related considerations, since Φ also induces an isomorphism of semigroup algebras determined by the partial ordering, as the l.c.m. algebra is (see the next section).

8. Semilattice algebras and the theorem of Lehmer-von Sterneck. We suppose now that (P,π) is an upper semilattice, i.e., $x \vee y$ exists for all $x,y \in P$. Then the vector space A_P is equipped with a product operation defined by (3.7), and the restriction to B_P^t (functions of the second variable only) is the semigroup algebra product of $(P,\vee)$. The l.c.m. product (2.5) and the max product (2.6) are examples of such semilattice algebra products.

Theorem (8.1) (Lehmer [49], von Sterneck [76]). Right matrix multiplication by ζ_π is an isomorphism from the algebra $(A_P,+,\vee)$ to the pointwise algebra $(A_P,+,\cdot)$.

The proof is an easy matrix computation. The isomorphism can be restricted to any of the pointwise subalgebras, such as CF_P, RF_P, I_π, B_P^t, which are (therefore) also semigroup (or $\vee$) algebras. The original theorem of von Sterneck was presented in the context of functions of one variable with the l.c.m. product. Lehmer's proof of von Sterneck's theorem was sufficiently general that it would work for any semigroup product xy on P such that $z \,\pi\, x$ and $z \,\pi\, y$ if and only if $z \,\pi\, xy$. Scheid [62] generalized Lehmer's result to a certain class of "partially defined semigroups" (see Section 2), using a variation of the argument which we will examine more closely below. Lehmer's theorem has also been rediscovered by Tainiter [79] for the case of finite semilattices, and in a somewhat different context. See Smith [73] for a generalization in a different direction.

Scheid's approach to the Lehmer-von Sterneck theorem is to map the semilattice algebra of functions of one variable (or, more generally, the algebra with tau product (2.9)) isomorphically to a subalgebra of the incidence algebra, rather than to the pointwise algebra. We assume that P has a smallest element 0 (which of course might have some other name in $\underline{N}$, as, e.g., 1 in the case of the divisibility ordering). Although we make no use of this fact, the existence of 0 implies that P is lattice. Define $T: B_P^t \to I_\pi$ by

$$(8.2) \qquad Tf(a,b) = \sum_{x \vee a = b} f(x) = \sum_x f(x)\,\delta(x \vee a, b).$$

Note that T is closely related to right multiplication by $\overline{\delta}$ induced by a closure operator (see Section 7), in that, for each fixed a, $x \to x \vee a$ is a closure operator on P. Since there is a different closure operator for each a, however, (8.2) is not simply the matrix product of f with a function in Λ_P.

Theorem (8.3) (Scheid [62]). T is an isomorphism from the semigroup algebra $(B_P^t, +, \vee)$ to a subalgebra of the incidence algebra $(I_\pi, +, *)$.

The proof is similar to that of (8.1), and the von Sterneck theorem is recovered by considering diagonal entries of the function Tf. The inverse of the map T is given by $T^{-1}g(x) = g(0,x)$. An interesting consequence of Scheid's theorem is the equivalence of the (generalized) Gegenbaur identity (6.15) with a (generalized) identity of von Sterneck:

$$\varphi_k \vee \varphi_j = \varphi_{k+j}. \tag{8.4}$$

Here we are considering the generalized totient functions as functions of one variable by setting $\varphi_k(x) = \varphi_k(0,x)$.

The proof (cf. Scheid [62]) is as follows. Write (6.15) in the equivalent form

$$\varphi_k \nu_j * \varphi_j = \varphi_{k+j}. \tag{8.5}$$

Evaluate (8.5) at (a,b) and multiply both sides by a factor of $\nu_{k+j}(0,a)$ to get

$$\sum_x \nu_k(0,a)\varphi_k(a,x)\nu_j(0,x)\varphi_j(x,b) = \nu_{k+j}(0,a)\varphi_{k+j}(a,b). \tag{8.6}$$

If we fix a, and consider the closure operator $x \vee a$, for which the set of closed elements is the set of y such that $a \,\pi\, y$, then it follows from (7.7) that

$$T\varphi_k(a,b) = \nu_k(0,a)\varphi_k(a,b). \tag{8.7}$$

Substitution of (8.7) into (8.6) yields

$$T\varphi_k * T\varphi_j = T\varphi_{k+j},$$

which implies (8.4).

Additional references for this section: Bell [11], Lehmer [50, 51], Scheid-Sivaramakrishnan [67], Stevens [77].

9. <u>The theorems of Lindström and H. J. S. Smith</u>. In this section we let P be a finite subset of $\underline{N}$ and π a lower semilattice ordering on P. For $f \in A_P$, define

$$\hat{f}(x,y) = f(x, x \wedge y), \quad x,y \in P. \tag{9.1}$$

Another straightforward matrix computation yields the following factorization of the matrices $\hat{f}$ into lower and upper triangular matrices:

$$f = (g\zeta_\pi^t) * \zeta_\pi, \quad \text{where } g = f * \mu_\pi. \tag{9.2}$$

Such a factorization makes it possible to compute $\det \hat{f}$ as the product of the diagonal entries of the factors on the right. In particular, if $P = \{x_1, \ldots, x_n\}$, then:

$$\det[f(x_i, x_i \wedge x_j)] = \prod_{i=1}^{n} f * \mu_\pi(x_i, x_i). \tag{9.3}$$

If we have n functions $f_i: P \to K$ and set $f(x_i, x_j) = f_i(x_j)$, then (9.3) becomes:

<u>Theorem</u> (9.4) (Lindström [52]).

$$\det\,[f_i(x_i \wedge x_j)] = \prod_{i=1}^{n} \sum_{j=1}^{n} f_i(x_j)\mu_\pi(x_j, x_i).$$

Note that in (9.3) and (9.4) it is unnecessary to assume that the labelling $x_1, \ldots, x_n$ of elements of P is in the natural order, or even that it refines the given partial order π, because a suitable change of basis will bring the elements into natural order without changing the value of the determinant.

Lindström arrived at (9.4) by generalizing a theorem about determinants on semilattices of sets under intersection. His principal application is to the construction of (± 1)-determinants of small order and large value, and no mention is made of arithmetic functions. The special case of (9.4) in which $P = \{1, 2, \ldots, n\}$ ordered by divisibility, and all the f_i are the same function f, is a much older theorem of H. J. S. Smith [74] (see also Dickson [31, pp. 122-128]):

(9.5) $$\det[f((i,j))] = F(1)F(2)\cdots F(n),$$

where $F = f * \mu$. In particular,

(9.6) $$\det((i,j)) = \varphi(1)\varphi(2)\cdots\varphi(n).$$

Returning to the case of an arbitrary semilattice $P = \{x_1,\ldots,x_n\}$, equipped with a map $\nu: P \to K^*$, by (9.3) we have for each k:

(9.7) $$\det[\nu_k(x_i,x_i \wedge x_j)] = \prod_{i=1}^{n} \varphi_k(x_i,x_i).$$

(This is non-trivial if we redefine $\nu(x,y) = \nu(y)/\nu(x)$ whether or not $x \pi y$, and $\varphi_k = \nu_k * \mu$.) Now suppose that $\nu(0) = 1$, where 0 is the smallest element of P. Let $k_1, k_2, \ldots, k_n$ be any sequence of natural numbers, and set $f(x_i,y) = \nu_{k_i}(0,y)$. Then (9.3) becomes:

(9.8) $$\det[\nu(x_i \wedge x_j)^{k_i}] = \prod_{i=1}^{n} \varphi_{k_i}(0,x_i).$$

Both the Smith determinant (9.6) and its unitary analogue (Jager [45]) are very special cases of (9.8).

Additional references: Carlitz [18], Smith [73].

10. <u>Generalized greatest-integer functions</u>. Let π_1 and π_2 be partial orders on $P_1 \subseteq P_2 \subseteq \underline{N}$ such that π_2 refines π_1. We write ζ_1, μ_1, τ_1, etc., for the appropriate elements of I_{π_1}, and similarly for π_2. Let $G_{12} = G = \zeta_1 * \zeta_2$. Such functions G are counting functions: $G(x,y)$ is the number of t in a π_2-interval $[x,y]$ such that $x \pi_1 t$. The simplest case is that in which $\pi_1 = \pi_2$ and $G = \tau$. A more important example is that in which π_1 is the divisibility ordering on $\underline{N}^*$ and π_2 is the total ordering. In this case

(10.1) $$G(x,y) = [y/x].$$

Another example (using relations which are not partial orders) appears in a recent paper of Apostol [2]: Let π_1 be defined on $\underline{N}^*$ by

$$x \; \pi_1 \; y \quad \text{if} \quad y/x \quad \text{is a k-free integer,}$$

and π_2 by

$$x \; \pi_2 \; y \quad \text{if} \quad y/x = \prod p_i^{\alpha_i}, \quad \text{each} \quad \alpha_i \not\equiv 0 \pmod{k},$$

where k is a fixed positive integer. Then π_2 refines π_1, so we may define $G = \zeta_1 * \zeta_2$. Apostol's "Möbius functions of order k" satisfy:

$$\mu_k = G * \mu, \tag{10.2}$$

where μ is the ordinary Möbius function.

If $P_1 = P_2$, then G is invertible in I_{π_2}, and

$$G^{-1} = \mu_2 * \mu_1. \tag{10.3}$$

In the case of (10.1), where μ_1 is the ordinary Möbius function and μ_2 is a difference operator, (10.3) is a theorem of Jacobsthal [44]. This result was the motivation for Carlitz [18], which paper was the source of the idea for the functions G.

We note some other immediate consequences.

$$\zeta_1 * f = G * F, \quad \text{where} \quad F = \mu_2 * f. \tag{10.4}$$

In particular,

$$\zeta_1 * \tau_2 = G * \zeta_2. \tag{10.5}$$

In the case of (10.1), (10.5) becomes

$$\sum_{z=x}^{y} \left[\frac{z}{x}\right] = \left(y - x + 1\right) + \cdots + \left(y - \left[\frac{y}{x}\right]x + 1\right),$$

from which one may deduce that, for any real number t and any $k \in \underline{N}^*$,

(10.6) $$\sum_{j=0}^{k-1} \left[t + \frac{j}{k}\right] = [kt].$$

Assuming a map $\nu: P_1 \to K^*$ and totient functions φ_k defined with respect to π_1 (i.e., $\varphi_k = \nu_k * \mu_1$), we have:

(10.7) $$\varphi_k * G = \nu_k * \zeta_2.$$

For the greatest-integer case (10.1), this is the Dirichlet formula (see Dickson [31, p. 150]):

(10.8) $$\sum_{j=1}^{n} \left[\frac{n}{j}\right] \varphi_k(j) = 1^k + 2^k + \cdots + n^k.$$

Now set $J_{12} = J = G_{12} * \mu_1$. In the case of (10.1), $J(x,y)$ is the number of positive integers $\le y/x$ and prime to x. If (P_1,π_1) is an upper semilattice, we define

(10.9) $$J^{(k)}(x_1,x_2,\ldots,x_k;y) = \sum_{z_1 \vee z_2 \vee \cdots \vee z_k = y} J(x_1,z_1)\cdots J(x_k,z_k).$$

It follows from the Lehmer-von Sterneck theorem (8.1) that

(10.10) $$\sum_{z\ \pi_1\ y} J^{(k)}(x_1,\ldots,x_k;z) = \prod_{i=1}^{k} G(x_i,y).$$

In the greatest-integer case, this is a theorem of von Sterneck,where (10.9) counts the number of sets $\{z_1, \ldots, z_k\}$ with g.c.d. prime to y and $z_j \le y/x_j$ for each j. Also note that:

(10.11) $$J * G = \zeta_1 * \tau_2 = G * \zeta_2.$$

This implies another result of von Sterneck:

(10.12) $$\sum_{k=1}^{n} J(m,k)\left[\frac{n}{k}\right] = \sum_{j=1}^{n} \left[\frac{j}{m}\right].$$

Let $T_{12} = T = \tau_1 * \zeta_2$. Then

(10.13) $$T = \zeta_1 * G,$$

which includes another identity of Dirichlet:

(10.14) $$\sum_{j=1}^{n} \tau(j) = \sum_{k=1}^{n} \left[\frac{n}{k}\right].$$

If $\sigma_k = \nu_k * \zeta_1$, then we have:

(10.15) $$\nu_k * G = \sigma_k * \zeta_2 = \varphi_k * T,$$

which incorporates identities of Dirichlet, Liouville, and Cesàro (see Dickson [31, pp. 282, 293]):

(10.16) $$\sum_{k=1}^{n} k\left[\frac{n}{k}\right] = \sum_{j=1}^{n} \sigma(j) = \sum_{m=1}^{n} \varphi(m)T\left[\frac{n}{m}\right].$$

The classical identities used as illustrations here all have in common the choice of divisibility for π_1 and the natural total order for π_2. (Of course, the identities all have unitary, Lucas, and other analogues.) The basic idea behind these identities could be improved on by taking (P_2, π_2) to be the reals with the usual ordering, or any partially ordered set in which (P_1, π_1) is imbedded <u>discretely</u>, in the following sense: given $x \in P_1$ and $y \in P_2$, there exist only finitely many $z \in P_1$ such that $x\ \pi_1\ z\ \pi_2\ y$. If P_1 is an upper semilattice, then any $y \in P_2$ which dominates an element of P_2 dominates a largest one which we may call $[y]$. For $f \in I_{\pi_1}$ and $g \in I_{\pi_2}$ (i.e., $g\colon P_2 \times P_2 \to K$ and $g(x,y) = 0$ unless $x\ \pi_2\ y$), it makes sense to write:

(10.17) $$f * g(x,y) = \sum_{z} f(x,z)g(z,y).$$

The sum is 0 unless $x \in P_1$, and has finitely many terms $\neq 0$ in any case. In the case of an upper semilattice,

(10.18) $$f * g(x,y) = \sum_{x \,\pi_1\, z \,\pi_1 [y]} f(x,z)g(z,y).$$

With this definition, we have an inversion formula,

(10.19) $$F = \zeta_1 * f \quad \text{if and only if} \quad f = \mu_1 * F,$$

which includes Hardy-Wright [41, p. 237, Thm. 268]. Similarly, one obtains Gupta [41, Thm. 4] for the generalized Möbius functions v_r (see Section 7).

11. Multiplicative and additive functions. It is clear that incidence functions on a partially ordered set (P,π) only reflect the local structure of (P,π), i.e., the structure of intervals. Thus it is appropriate to describe P in terms of local properties it does or does not have. We say that P is a local lattice if every interval is a lattice. Similarly, P is locally Boolean (respectively, locally distributive, locally modular) if every interval is a Boolean algebra (respectively, distributive lattice, modular lattice). The unitary ordering (5.3) on $\underline{N}^*$ is an example of a locally Boolean local lattice which is not a lattice. All of the orderings (5.1)-(5.4) are locally distributive.

Let (P,π) be a local lattice. An invertible function $f \in I_\pi$ is called multiplicative (or factorable) if

(11.1) $$f(x \vee z, y \vee w) = f(x,y)f(z,w)$$

for all x, y, z, w in any given interval such that $x \le y$, $z \le w$, and $x \wedge z = y \wedge w$. Similarly, one defines additive functions by replacing the product in (11.1) with a sum.

This definition mimics the usual definition for arithmetic functions, and indeed reduces to it for functions in the Dirichlet algebra I_D. The notion of factorability is extremely useful in the study of arithmetic functions because (a) factorable functions are completely determined by their values on prime powers, and (b) they form a group under Dirichlet product which contains a lot of interesting and important functions. One might hope to obtain similar benefits from a study of factorability on local lattices in general, but the definition we have given is really useful only in the locally distributive case. We recall some results from the author's papers [69, 70, 71].

Theorem (11.2). *δ_π is multiplicative if and only if (P,π) is locally modular.*

Obviously the definition cannot be very useful if the identity element of the algebra fails to be multiplicative. The only obviously multiplicative function under all circumstances is ζ_π, and the real test of usefulness is whether its powers are also multiplicative.

Theorem (11.3). *The following are equivalent:*

(a) *μ_π is multiplicative.*

(b) *τ_π is multiplicative.*

(c) *The multiplicative functions form a group under $*$.*

(d) *(P,π) is locally distributive.*

Theorem (11.4). *Suppose that (P,π) satisfies the Jordan-Dedekind chain condition. Then the following are equivalent:*

(a) *d and λ are additive.*

(b) *d and λ' are additive.*

(c) *(P,π) is locally distributive.*

(See (6.2)-(6.4) for definition of d, λ, λ'.)

We now suppose that (P,π) is locally distributive. A function $f \in I_\pi$ is called *translation invariant* if

(11.5) $$f(x,y) = f(x \vee z, y \vee z),$$

whenever x, y, z are elements of an interval such that $x \leq y$ and $x \wedge z = y \wedge z$. An equivalent definition is:

(11.6) $$f(x \wedge y, x) = f(y, x \vee y)$$

for all x, y in any interval.

Theorem (11.7). *An invertible function f is multiplicative if and only if f is translation invariant and*

(11.8) $$f(x, y \vee z) = f(x,y)f(x,z)$$

for any x, y, z in an interval such that $x = y \wedge z$.

<u>Theorem</u> (11.9) <u>If</u> f <u>is completely multiplicative (3.9) and translation invariant, then</u> f <u>is multiplicative</u>.

The earliest attempt to define factorability in the incidence algebra goes back to Ward [82], who used the condition

$$(11.10) \qquad f(x,y)f(z,w) = f(x \wedge z, y \wedge w) f(x \vee z, y \vee w),$$

for elements x, y, z, w in a lattice (or in an interval of a local lattice). He was able to show that if τ satisfies (11.10) then (P,π) is locally distributive. The inadequacy of (11.10) was demonstrated by Bender [13], who showed that the functions satisfying (11.10) are closed under $*$ if and only if (P,π) is locally Boolean.

For certain problems outside the area of arithmetic functions, it would be highly desirable to have a satisfactory theory of multiplicative functions for other classes of local lattices, such as locally modular or locally geometric. In this connection, Doubilet, Rota, and Stanley [32] have introduced such a notion for partition lattices.

Another problem of interest would be the determination of the subspace of I_π spanned by the multiplicative functions. Even in the Dirichlet case I_D, so far as we know, the only results in this direction are those of Carlitz [22].

12. <u>The logarithmic operator</u>. In this section we summarize some results from the author's paper [70]. We assume that the base field K is the real field $\underline{R}$. Let (P,π) be a partial ordering of a subset of $\underline{N}$, and let ν be a one-to-one, increasing mapping from P into the positive reals $\underline{R}^+$, with the usual ordering. As before, ν is also the name of an incidence function defined by $\nu(x,y) = \nu(y)/\nu(x)$. Let ρ be an order-compatible relation such that I_ρ is commutative and $\nu \in I_\rho$. Let G_ρ be the group of units of I_ρ and G_ρ^+ the subgroup of functions f such that $f(x,x) > 0$ for all $x \in P$. (It follows from Theorem (5.10) that G_ρ is the intersection of I_ρ with the group of units of I_π.) For functions $f \in G_\rho^+$, we define the logarithmic operator L:

$$(12.1) \qquad Lf(x,x) = \log f(x,x) ,$$

and if $x \neq y$,

(12.2) $$Lf(x,y) = \sum_{x \,\pi\, z \,\pi\, y} f \log \nu(x,z) f^{-1}(z,y).$$

Theorem (12.3). L is a group isomorphism from $(G^+_\rho,*)$ to $(I_\rho,+)$.

For each real number r, we define $f^r = L^{-1}(r\,L\,f)$, for $f \in G^+_\rho$. The usual laws of exponents hold, and $f \to f^r$ is an automorphism of G^+_ρ.

We now assume further that (P,π) is a locally distributive local lattice, that ν is multiplicative, and that ρ is translation invariant, in the sense that $[x,y]\ \rho\ [x \vee z,\ y \vee z]$ if $x \wedge z = y \wedge z$. Let M_ρ denote the set of multiplicative functions in I_ρ. M_ρ is a group, because it is the intersection of G^+_ρ with the group of multiplicative functions in I_π. An interval in P is called narrow if no element in it has a complement, except for the end points.

Theorem (12.4). $h \in L(M_\rho)$ if and only if h is translation invariant and is 0 except on narrow intervals.

The restriction of L to M_ρ is an isomorphism between $(M_\rho,*)$ and $(L(M_\rho),+)$.

Theorem (12.5). If there is a bijection between ρ-classes of narrow intervals and ρ-classes of all intervals, then $(M_\rho,*) \cong (I_\rho,+)$.

These results generalize and unify the arithmetic function results of Rearick [57], which was the source of the basic idea. They also include the more general results of Goldsmith [39].

In a recent paper, Fredman [33] has introduced a different type of logarithmic operator on arithmetic functions. This could also be done for incidence functions of the form $\delta + f$, where $f(x,x) = 0$ for all x. It would be of interest to know what the relationship to the Rearick-type operator is, if any.

13. Generalizations of classical identities. The usefulness of the notion of factorability given in Section 11 was demonstrated by the author in [71], in which relatively simple proofs were given of generalizations of the identities of Brauer-Rademacher [16], Hölder [42], Landau [47], and others. These generalized identities include as special cases most of the generalizations of the same identities that have been published by others, and include unitary and Lucas analogues as well. We summarize the principal results of [71], but in a slightly less general form than is given there, for reasons of economy and clarity.

We assume that (P,π) is a lower semilattice in which each interval is a direct product of a finite number of finite chains. (The partial orderings (5.1)-(5.4) are all of this sort, and any such partial ordering is locally distributive.) As in Section 6, we assume the existence of a map $\nu: P \to K^*$, but we insist that ν be one-to-one and that the corresponding incidence function ν defined by (6.6) be multiplicative. By Theorem (11.9), this will be the case if ν is translation invariant. In particular, this will be the case if ν satisfies the condition of [71], namely, that it takes its values in $\underline{N}^*$ and that it takes infima and suprema (where they exist) to g.c.d.'s and l.c.m.'s, respectively.

Since every interval is a product of chains, the values of any multiplicative function are determined by its values on intervals which are chains. We will say that a function $f \in M_\pi$ (the group of multiplicative functions in I_π) is constant on chains if its value on any chain of length ≥ 2 is the same as its value on any 2-element subchain, and the latter is not zero. If f is constant on chains, a companion function for f is any function $h \in M_\pi$ such that $h(x,y) = f(x,y) - 1$ whenever y covers x. (Such functions always exist; e.g., $h = f * \mu$ is such.)

Theorem (13.1). Let f be a constant on chains, h a companion function, and x, y, $t \in P$ such that $x \,\pi\, (y \wedge t)$. Then:

$$\sum_{\substack{x \,\pi\, z \,\pi\, y \\ t \wedge z = x}} f(x,z)\mu(z,y) = \mu(x,y)\mu(y \wedge t,y)h(y \wedge t,y), \tag{13.2}$$

$$(13.3)\qquad \sum_{\substack{x \,\pi\, z \,\pi\, y \\ t \wedge z = x}} h(x,z)\,\mu(x,z)^2 = f(x,y)/f(x,y \wedge t),$$

$$(13.4)\qquad \sum_{x \,\pi\, z \,\pi\, (y \wedge t)} \frac{h(z,y)\,\mu(z,y)}{f(z,y)} = \frac{h(y \wedge t,y)\,\mu(y \wedge t,y)}{f(x,y)},$$

$$(13.5)\qquad \frac{1}{f(x,y)} \sum_{\substack{x \,\pi\, z \,\pi\, y \\ t \wedge z = x}} f(x,z)\,\mu(z,y) = \mu(x,y) \sum_{x \,\pi\, z \,\pi\, (y \wedge t)} \frac{h(z,y)\mu(z,y)}{f(z,y)}$$

<u>If, in addition,</u> $h(r,s) \neq 0$ <u>whenever</u> $r \,\pi\, s$, <u>then</u>:

$$(13.6)\qquad \sum_{\substack{x \,\pi\, z \,\pi\, y \\ t \wedge z = x}} f(x,z)\,\mu(z,y) = (x,y)h(x,y) \sum_{x \,\pi\, z \,\pi\, (y \wedge t)} \frac{f(x,z)\,\mu(z,y)}{h(x,z)}.$$

These identities are generalizations of identities which may be attributed, respectively, to Subbarao, Cohen-Landau, Anderson-Apostol-Hölder, Brauer-Rademacher, and Cohen.

The more classical forms of these identities are obtained by setting

$$(13.7)\qquad f(x,y) = \frac{\nu_k(x,y)}{\varphi_{k,j}(x,y)}, \qquad h(x,v) = \frac{\nu_j(x,y)}{\varphi_{k,j}(x,y)} .$$

The verification that f, h satisfy the hypotheses of the theorem rests on the assumptions that ν is one-to-one and translation invariant. In particular, for the case of $k = 1$, $j = 0$, the identities become:

$$(13.8)\qquad \sum_{\substack{x \,\pi\, z \,\pi\, y \\ t \wedge z = x}} \frac{\nu}{\varphi}(x,z)\,\mu(z,y) = \mu(x,y)\frac{\mu}{\varphi}(y \wedge t,y),$$

$$(13.9)\qquad \sum_{\substack{x \,\pi\, z \,\pi\, y \\ t \wedge z = x}} \frac{\mu^2}{\varphi}(x,z) = \frac{\varphi(x,y \wedge t)\,\nu(y \wedge t,y)}{\varphi(x,y)},$$

$$(13.10)\qquad \sum_{x \pi z \pi (y \wedge t)} \nu(x,z)\mu(z,y) = \varphi(x,y)\frac{\mu}{\varphi}(y \wedge t,y),$$

$$(13.11)\qquad \sum_{\substack{x \pi z \pi y \\ t \wedge z = x}} \frac{\nu}{\varphi}(x,z)\mu(z,y) = \mu(x,y) \sum_{x \pi z \pi (y \wedge t)} \nu(x,z)\mu(z,y).$$

(Both (13.5) and (13.6) yield (13.11) in this case.)

Other sets of interesting identities result from setting:

$$(13.12)\qquad f(x,y) = \frac{\varphi_{k,j}(x,y)}{\nu_k(x,y)}, \quad h(x,y) = \frac{\mu(x,y)}{\nu_{k-j}(x,y)},$$

or:

$$(13.13)\qquad f(x,y) = 2^{\lambda(x,y)}, \quad h = \zeta.$$

These are written out in detail in [71], and will not be repeated. here.

REFERENCES

1. D. R. Anderson and T. M. Apostol, The evaluation of Ramanujan's sum and its generalizations, Duke Math. J. 20 (1953), 211-216.

2. T. M. Apostol, Möbius functions of order k, Pacific J. Math. 32 (1970), 21-27.

3. T. M. Apostol, Some properties of completely multiplicative arithmetical functions, Amer. Math. Monthly 78 (1971), 266-271.

4. E. T. Bell, An arithmetical theory of certain numerical functions, Univ. of Washington Publications in Mathematics 1 (1915), No. 1.

5. E. T. Bell, On a certain inversion in the theory of numbers, Tôhoku Math. J. 17 (1920), 221-231.

6. E. T. Bell, Extension of Dirichlet multiplication and Dedekind inversion, Bull. Amer. Math. Soc. 28 (1922), 111-122.

7. E. T. Bell, Euler algebra, Trans. Amer. Math. Soc. 25 (1923), 135-154.

8. E. T. Bell, Algebraic Arithmetic, Colloq. Publ. Vol. VII, Amer. Math. Soc., New York, 1927.

9. E. T. Bell, Transformations of relations between numerical functions, Ann. Mat. Pura Appl. (Ser. IV) 4 (1927), 1-6.

10. E. T. Bell, An outline of a theory of arithmetic functions, J. Indian Math. Soc. 17 (1928), 249-260.

11. E. T. Bell, Functional equations for totients, Bull. Amer. Math. Soc. 37 (1931), 85-90.

12. E. T. Bell, Factorability of numerical functions, Bull. Amer. Math. Soc. 37 (1931), 251-253.

13. E. A. Bender, Numerical identities in lattices with an application to Dirichlet products, Proc. Amer. Math. Soc. 15 (1964), 8-13.

14. M. G. Beumer, The arithmetic function $\tau_k(n)$, Amer. Math. Monthly 69 (1962), 777-781.

15. G. Birkhoff, Lattice Theory, Third Edition, Colloq. Publ. Vol. XXV, Amer. Math. Soc., Providence, 1967.

16. A. Brauer and H. Rademacher, Aufgabe 31, Jber. Deutsch. Math. - Verein 35 (1926), 94-95 (suppl.)

17. R. G. Buschman, Identities involving products of number-theoretic functions, Proc. Amer. Math. Soc. 25 (1970), 307-309.

18. L. Carlitz, Some matrices related to the greatest integer function, J. Elisha Mitchell Scientific Society 76 (1960), 5-7.

19. L. Carlitz, Rings of arithmetic functions, Pacific J. Math. 14 (1964), 1165-1171.

20. L. Carlitz, Arithmetic functions in an unusual setting, I, Amer. Math. Monthly 73 (1966), 582-590.

21. L. Carlitz, Arithmetic functions in an unusual setting, II, Duke Math. J. 34 (1967), 757-760.

22. L. Carlitz, Sums of arithmetic functions, Collect. Math. 20 (1969), 107-126.

23. E. D. Cashwell and C. J. Everett, The ring of number theoretic functions, Pacific J. Math. 9 (1959), 975-985.

24. E. Cohen, Representations of even functions (mod r), I. Arithmetical identities, Duke Math. J. 25 (1958), 401-422.

25. E. Cohen, Representations of even functions (mod r), II. Cauchy products, Duke Math. J. 26 (1959), 165-182.

26. E. Cohen, The Brauer-Rademacher identity, Amer. Math. Monthly 67 (1960), 30-33.

27. E. Cohen, Arithmetical inversion formulas, Canad. J. Math. 12 (1960), 399-409.

28. E. Cohen, Arithmetical functions associated with the unitary divisors of an integer, Math. Z. 74 (1960), 66-80.

29. E. Cohen, Unitary products of arithmetical functions, Acta Arith. 7 (1961), 29-38.

30. T. M. K. Davison, On arithmetic convolutions, Canad. Math. Bull. 9 (1966), 287-296.

31. L. E. Dickson, History of the Theory of Numbers, Vol. 1, Carnegie Institution, Washington, 1919.

32. P. Doubilet, G.-C. Rota, and R. Stanley, The idea of generating function, M.I.T. preprint, Feb. 1971.

33. M. L. Fredman, Arithmetical convolution products and generalizations, Duke Math. J. 37 (1970), 231-242.

34. M. D. Gessley, A generalized arithmetic composition, Amer. Math. Monthly 74 (1967), 1216-1217.

35. A. A. Gioia, The K-product of arithmetic functions, Canad. J. Math. 17 (1965), 970-976.

36. D. L. Goldsmith, On the multiplicative properties of arithmetic functions, Pacific J. Math. 27 (1968), 283-304.

37. D. L. Goldsmith, A generalization of some identities of Ramanujan, Rend. Mat. (Ser. VI) 2 (1969), 473-479.

38. D. L. Goldsmith, A note on sequences of almost multiplicative arithmetic functions, Rend. Mat. (Ser. VI) 3 (1970), 167-170.

39. D. L. Goldsmith, A generalized convolution for arithmetic functions, Duke Math. J. 38 (1971), 279-283.

40. H. Gupta, A generalization of the Möbius function, Scripta Math. 19 (1953), 121-126.

41. G. H. Hardy and E. M. Wright, An Introduction to the Theory of Numbers, Third Edition, The Clarendon Press, Oxford, 1954.

42. O. Hölder, Zur Theorie der Kreisteilungsgleichung $K_m(x) = 0$, Prace Mat. 43 (1936), 13-23.

43. L. C. Hsu, Abstract theory of inversion of iterated summations, Duke Math. J. 14 (1947), 465-473.

44. E. Jacobsthal, Über die grösste ganze Zahl, II, Norske Vid. Selsk. Forh. (Trondheim) 30 (1957), 6-13.

45. H. Jager, The unitary analogues of some identities for certain arithmetical functions, Nederl. Akad. Wetensch. Proc. Ser. A. 64 (= Indag. Math. 23) (1961), 508-515.

46. V. L. Klee, A generalization of Euler's φ-function, Amer. Math. Monthly 55 (1948), 358-359.

47. E. Landau, Über die zahlentheoretische Funktion $\varphi(m)$ und ihre Beziehung zum Goldbachschen Satz, Göttinger Nachr. (1900), 177-186.

48. D. H. Lehmer, On the r-th divisors of a number, Amer. J. Math. 52 (1930), 293-304.

49. D. H. Lehmer, On a theorem of von Sterneck, Bull. Amer. Math. Soc. 37 (1931), 723-726.

50. D. H. Lehmer, A new calculus of numerical functions, Amer. J. Math. 53 (1931), 843-854.

51. D. H. Lehmer, Arithmetic of double series, Trans. Amer. Math. Soc. 33 (1931), 945-952.

52. B. Lindström, Determinants on semilattices, Proc. Amer. Math Soc. 20 (1969), 207-208.

53. P. J. McCarthy, The generation of arithmetical identities, J. Reine Angew. Math. 203 (1960), 55-63.

54. P. J. McCarthy, Some remarks on arithmetical identities, Amer. Math. Monthly 67 (1960), 539-548.

55. P. J. McCarthy, Some more remarks on arithmetical identities, Portugal. Math. 21 (1962), 45-57.

56. W. Narkiewicz, On a class of arithmetical convolutions, Colloq. Math. 10 (1963), 81-94.

57. D. Rearick, Operators on algebras of arithmetic functions, Duke Math. J. 35 (1968), 761-766.

58. D. Rearick, The trigonometry of numbers, Duke Math. J. 35 (1968), 767-776.

59. G. C. Rota, On the foundations of combinatorial theory, I. Theory of Möbius functions, Z. Wahrscheinlichkeitstheorie und Verw. Gebiete 2 (1964), 340-368.

60. H. Scheid, Arithmetische Funktionen über Halbordnung, I, J. Reine Angew. Math. 231 (1968), 192-214.

61. H. Scheid, Arithmetische Funktionen über Halbordnung, II, J. Reine Angew. Math. 232 (1968), 207-220.

62. H. Scheid, Einige Ringe zahlentheoretischer Funktionen, J. Reine Angew. Math. 237 (1969), 1-11.

63. H. Scheid, Über ordnungstheoretische Funktionen, J. Reine Angew. Math. 238 (1969), 1-13.

64. H. Scheid, Funktionen über lokal endlichen Halbordnung, I, Monatsh. Math. 74 (1970), 336-347.

65. H. Scheid, Funktionen über lokal endlichen Halbordnung, II, Monatsh. Math. 75 (1971), 44-56.

66. H. Scheid, Über die Möbiusfunktion einer lokal endlichen Halbordnung (to appear).

67. H. Scheid and R. Sivaramakrishnan, Certain classes of arithmetic functions and the operation of additive convolution, J. Reine Angew. Math. 245 (1970), 201-207.

68. R. Sivaramakrishnan, The arithmetic function $\tau_{k,r}(n)$, Amer. Math. Monthly 75 (1968), 988-989.

69. D. A. Smith, Incidence functions as generalized arithmetic functions, I, Duke Math. J. 34 (1967), 617-634.

70. D. A. Smith, Incidence functions as generalized arithmetic functions, II, Duke Math. J. 36 (1969), 15-30.

71. D. A. Smith, Incidence functions as generalized arithemtic functions, III, Duke Math. J. 36 (1969), 353-368.

72. D. A. Smith, Multiplication operators on incidence algebras, Indiana Univ. Math. J. 20 (1970), 369-383.

73. D. A. Smith, Bivariate function algebras on posets, J. Reine. Angew. Math. (to appear).

74. H. J. S. Smith, On the value of a certain arithmetical determinant, Proc. London Math. Soc. 7 (1875-6), 208-212 (Collected Mathematical Papers, vol. II, p. 161).

75. R. P. Stanley, Structure of incidence algebras and their automorphism groups, Bull. Amer. Math. Soc. 76 (1970), 1236-1239.

76. R. Daublebsky von Sterneck, Ableitung zahlentheoretischer Relationen mit Hilfe eines mehrdimensionalen Systems von Gitterpunkten, Monatsh. Math. 5 (1894), 255-266.

77. H. Stevens, Generalizations of the Euler φ-function, Duke Math. J. 38 (1971), 181-186.

78. M. V. Subbarao, The Brauer-Rademacher identity, Amer. Math. Monthly 72 (1965), 135-138.

79. M. Tainiter, Generating functions on idempotent semigroups with applications to combinatorial analysis, J. Combinatorial Theory 5 (1968), 273-288.

80. M. Tainiter, Incidence algebras on generalized semigroups, I, Brookhaven preprint BNL13653, April 1969 (to appear in J. Combinatorial Theory).

81. R. Vaidyanathaswamy, The theory of multiplicative arithmetic functions, Trans. Amer. Math. Soc. 33 (1931), 579-662.

82. M. Ward, The algebra of lattice functions, Duke Math. J. 5 (1939), 357-371.

83. L. Weisner, Abstract theory of inversion of finite series, Trans. Amer. Math. Soc. 38 (1935), 474-484.

ON SOME ARITHMETIC CONVOLUTIONS

M. V. Subbarao, University of Alberta

0. Introduction. In this paper we first review some of the known arithmetical convolutions with particular reference to a class of convolutions which may be called Lehmer's ψ-products. These products are general enough to include as special cases the well known Dirichlet and unitary products and other Narkiewicz-type products. We indicate a few new cases of ψ-products, and in particular study in some detail a new convolution, called "exponential convolution", which is a variant of Lehmer's ψ-product.

For arbitrary arithmetic functions α, β, this is defined by the equations

$$(\alpha \odot \beta)(1) = \alpha(1)\beta(1),$$

and if $n > 1$ has the canonical form

$$n = p_1^{a_1} \cdots p_r^{a_r},$$

then

$$(\alpha \odot \beta)(n) = \prod_{\substack{b_j c_j = a_j \\ j = 1,\ldots,r}} \alpha(\pi p_j^{b_j})\beta(\pi p_j^{c_j}).$$

This is unlike most known arithmetical convolutions. For example, it is not of the form

$$\sum_{d|n} \alpha(d)\beta(n/d)$$

or of the form

$$\sum_{a \le n} \alpha(a)\beta(n-a).$$

Yet it is commutative and associative. For $n > 1$, among the divisors over which it is summed, the smallest is not 1, but the core of n, namely the product of the distinct prime factors of n.

We obtain some of the simplest properties associated with this convolution. For example, calling d an exponential divisor of $n = p_1^{a_1} \cdots p_r^{a_r}$ if $d = p_1^{b_1} \cdots p_r^{b_r}$ where $b_j | a_j$, and denoting the number of such divisors by $\tau^{(e)}(n)$, we obtain fairly satisfactory results for the order of $\tau^{(e)}(n)$. For example,

$$\overline{\lim_{n\to\infty}} \log \tau^{(e)}(n) \log\log n/\log n = \tfrac{1}{2}\log 2.$$

Analogous to the Dirichlet divisor problem, we have here the corresponding divisor problem for exponential divisors, namely to find the exact order of the error function for the summatory function

$$T(x) = \sum_{n \le x} \tau^{(e)}(n).$$

We can show that $T(x) = Ax + E(x)$ where $E(x) = O(x^{\frac{1}{2}+\epsilon})$ for every positive ϵ. But the exact order of $E(x)$ is still an open question.

The commutative semigroup of arithmetic functions defined by the exponential convolution $\odot$ has zero divisors. Whether the sub-semigroup formed by the non-zero-divisors has the unique factorization property is another of the unsolved problems associated with this convolution. Some other problems appear in the last section.

1. Definitions and Notations. By an arithmetic function $\alpha(n)$ we mean a complex-valued function defined for all positive integers n (and in some cases for $n = 0$ also). We denote the set of positive integers by Z, the set of arithmetic functions by S, and arbitrary arithmetic functions by α, β, γ. We write $n, m, a, b, c, a_1, \ldots, a_r, b_1, \ldots, b_r, c_1, \ldots, c_r$ to mean always positive integers, while $h_1, h_2, \ldots$ represent non-negative integers. We denote the sequence of all primes by $q_1, q_2, \ldots,$ so that $q_1 = 2, q_2 = 3,$ etc. Also $p_1, p_2, \ldots, p_r$ denote arbitrary primes. If $n > 1$, its canonical form is always assumed to be

$$n = p_1^{a_1} \cdots p_r^{a_r}. \tag{1.1}$$

Note that every $n \geq 1$ has the unique representation

(1.2) $$n = q_1^{h_1} q_2^{h_2} \cdots,$$

where all but a finite number of the h's are zero.

Let (a,b) and $[a,b]$ denote, respectively, the g. c. d. and l. c. m. of a and b.

An arithmetic function α is said to be <u>multiplicative</u> if

$$\alpha(ab) = \alpha(a)\alpha(b)$$

for all a, b such that $(a,b) = 1$. The notion of multiplicativity can be generalized or specialized in many ways.

An illustration of each kind is provided by the following definitions (which we require later on).

An arithmetic function $\alpha(n)$ is said to be

(1.3) <u>semi-multiplicative</u> [29] if for all a and b,

$$\alpha(a)\alpha(b) = \alpha((a,b))\alpha([a,b]);$$

(1.4) <u>exponentially multiplicative</u> if α is multiplicative and, whenever $(a,b) = 1$,

$$\alpha(p^{ab}) = \alpha(p^a)\alpha(p^b)$$

for all primes p.

There are several other classes of arithmetic functions including completely multiplicative, completely unmultiplicative and almost multiplicative (Goldsmith [22, 23]).

2. <u>Some types of convolutions</u>. Given two arithmetic functions α and β, their sum (also called <u>natural sum</u>) is defined by

$$(\alpha + \beta)(n) = \alpha(n) + \beta(n) \qquad (n \in Z).$$

However, their product may be defined in several ways, thus giving rise to different types of arithmetical convolutions. Among the simplest and most widely known are the following three:

(2.1) <u>Natural product</u> $\alpha\beta$ (also written as $\alpha \times \beta$) defined by

$$(\alpha\beta)(n) = \alpha(n)\beta(n);$$

(2.2) <u>Dirichlet product</u> $\alpha \cdot \beta$ given by

$$(\alpha \cdot \beta)(n) = \sum_{ab=n} \alpha(a)\beta(b);$$

(2.3) <u>Unitary product</u> $\alpha * \beta$ defined by

$$(\alpha * \beta)(n) = \sum_{\substack{ab=n \\ (a,b)=1}} \alpha(a)\beta(b).$$

The theory of arithmetic functions connected with Dirichlet convolutions was first investigated by E. T. Bell [1, 2, 3, 4], and later extensively by R. Vaidyanathaswamy [42, 43], who also introduced the convolution now known as unitary product. This convolution was later extensively studied, among others, by Eckford Cohen [12]. The ring $(S,+,\cdot)$ has the unique factorization property, as was shown by Cashwell and Everett [8], but the ring $(S,+,*)$ is not even an integrity domain.

(2.4) <u>The l. c. m. product</u> $\alpha \oplus \beta$ of α and β is defined by

$$(\alpha \oplus \beta)(n) = \sum_{[a,b]=n} \alpha(a)\beta(b).$$

This convolution was studied in detail by D. H. Lehmer [25]. In view of Von Sterneck's theorem [24, p. 955] that if $\gamma = \alpha \oplus \beta$, then

$$\left(\sum_{a|n} \alpha(a)\right)\left(\sum_{a|n} \beta(a)\right) = \left(\sum_{a|n} \gamma(a)\right), \tag{2.5}$$

the calculation of l. c. m. products is essentially reduced to that of Dirichlet and natural products, since if μ is the Möbius function, we have from (2.5),

$$\gamma = (\alpha \cdot E)(\beta \cdot E) \cdot \mu \tag{2.6}$$

where E is the arithmetical function defined by

$$E(n) = 1 \quad \text{for all } n \in Z. \tag{2.7}$$

(2.8) <u>The Cauchy product</u> of α and β is given by

$$\sum_{\substack{a+b=n \\ a \geq 0}} \alpha(a)\beta(b) \qquad (n \in Z).$$

We here require that $\alpha(n)$ and $\beta(n)$ be defined for $n = 0$ also.

The Cauchy product was studied by several authors including E. T. Bell [1] and Eckford Cohen [10, 11]. The set S with natural sum and Cauchy product is an integrity domain in which there is essentially a single prime α defined by $\alpha(1) = 1$, $\alpha(n) = 0$ $(n \neq 1)$.

(2.9) <u>The Lucas-Carlitz product</u>. Recently, L. Carlitz [6, 7] introduced this product which is analogous to the Cauchy product. Let p be a fixed prime and put

$$n = n_0 + n_1 p + n_2 p^2 + \cdots \quad (0 \leq n_j < p),$$

$$r = r_0 + r_1 p + r_2 p^2 + \cdots \quad (0 \leq r_j < p).$$

Then, a result dating back to Lucas (see [6], p. 583) states that

(2.10) $$\binom{n}{r} \equiv \binom{n_0}{r_0}\binom{n_1}{r_1}\binom{n_2}{r_2}\cdots \pmod{p}.$$

Hence the binomial coefficient $\binom{n}{r}$ is relatively prime to p if and only if

(2.11) $$0 \leq r_j \leq n_j \qquad (j = 0, 1, 2, \ldots).$$

Using this fact, Carlitz defines the new product of α and β by the expression

$$\sum \alpha(a)\beta(n-a),$$

where the sum is restricted to those a that satisfy (2.11). (Here $\alpha(0)$, $\beta(0)$ are assumed to be defined.) Carlitz calls this the Lucas product of α and β. However, since Lucas did not even think of a convolution based on (2.10), it would be more appropriate to call this the the <u>Lucas-Carlitz</u> product. Carlitz has an unproved conjecture regarding the zero divisors of the ring constituted by S, natural addition and Lucas-Carlitz multiplication [6].

(2.12) Narkiewicz's convolution [28]. For each positive n, let A_n be a non-empty set of positive divisors of n. For each $\alpha, \beta \in S$, the product $\alpha \circ \beta$ is defined by Narkiewicz by the relation

$$(\alpha \circ \beta)(n) = \sum_{\substack{ab=n \\ a \in A_n}} \alpha(a)\beta(b).$$

Narkiewicz gives conditions on A_n under which this convolution is commutative and associative and shows that the semigroup $(S, \circ)$ has an identity element if and only if $\{1,n\} \subseteq A_n$ for every $n \in Z$, and then the identity element is the arithmetic function η defined by

$$\eta(n) = \begin{cases} 1, & n = 1, \\ 0, & n > 1. \end{cases} \qquad (2.13)$$

Further, Narkiewicz shows that the units of $(S, \circ)$ are those $\alpha \in S$ for which $\alpha(1) \neq 0$. Also, the convolution $\circ$ preserves multiplicativity (that is, $\alpha \circ \beta$ is multiplicative whenever α and β are) if and only if

$$A_{mn} = A_m \times A_n \qquad (m, n \in Z),$$

where $A_m \times A_n$ represents the set

$$\{ab: a \in A_m, \ b \in A_n\}.$$

All these properties hold of course in the case of Dirichlet and unitary convolutions.

There are some unsolved problems about this convolution such as in [28, p. 87].

The Narkiewicz product is not general enough to include the convolutions defined by the l. c. m., Cauchy, Lucas-Carlitz or natural products.

(2.14) The k-product (Gioia and Subbarao [19, 38, 39]) of α and β is defined by

$$\sum_{ab=n} \alpha(a)\beta(b)k((a,b))$$

where the function k satisfies the condition for associativity, namely,

(2.15) $$k((a,b))k((ab,c)) = k((a,bc))k((b,c))$$

for all $a, b, c \in Z$. The commutativity of the product is automatic. The k-product convolution appears to be the first generalization in the literature involving a kernel. It extends in an elegant manner many of the nice properties and identities associated with Dirichlet and unitary products. We refer to [19, 38, 39] for details.

The k-product is further generalized by T. M. K. Davison [15] as follows.

(2.16) Davison's product of α and β is given by

$$\sum_{ab=n} \alpha(a)\beta(b)A(a,b)$$

where $A(a,b)$ is a function of the two variables a and b, instead of being a function of their g. c. d. as in (2.14). See also Gesely [18].

(2.17) Remark. It is possible to construct a variety of interesting special cases of the products (2.12), (2.14) and (2.16). As examples which should be worthy of a detailed study, we mention the following.

(2.18) The Semi-unitary product of α and β may be defined by

$$\sum_{\substack{ab=n \\ (a,b)_*=1}} \alpha(a)\beta(b).$$

Here $(a,b)_*$, called the semi-unitary g. c. d. of a to b, is the largest divisor of a which is a unitary divisor of b. (c is called a unitary divisor of b if $c|b$ and $(c,b/c) = 1$.)

(2.19) Bi-unitary product of α and β, denoted by $\alpha{*}{*}\beta$, may be defined by

$$\sum_{\substack{ab=n \\ (a,b)_{**}=1}} \alpha(a)\beta(b),$$

where $(a,b)_{**}$ denotes the largest positive integer which is a unitary divisor of both a and b.

In fact some divisor functions related to these convolutions are already considered respectively by Chidambaraswamy [9] and Suryanarayana [41].

Some interesting convolutions that await detailed study, all special cases of (2.12) and (2.16), are defined by the following products:

$$\sum_{\substack{ab=n \\ \gamma(a)=\gamma(n)}} \alpha(a)\beta(b); \tag{2.20}$$

$$\sum_{\substack{ab=n \\ \gamma(a)=\gamma(b)}} \alpha(a)\beta(b); \tag{2.21}$$

$$\sum_{\substack{ab=n \\ (a,b)_k^*=1}} \alpha(a)\beta(b). \tag{2.22}$$

Here $\gamma(a)$, the core of a, denotes the product of the distinct prime factors of a; and $(a,b)_k^*$ denotes the greatest k-th power divisor of a which is a unitary divisor of b. (See [34].)

We shall not refer to several other convolutions, in the literature, such as those associated with the work of L. Weisner [44], G. C. Rota [32], H. H. Crapo [14], D. A. Smith [35, 36, 37], D. L. Goldsmith [22, 23], as well as E. Cohen's convolution of arithmetic functions of finite abelian groups [13]. It should be pointed out that the bibilography at the end of the paper gives only an illustrative list of some of the work done on convolutions.

3. <u>The Lehmer ψ-product $\alpha \bigcirc \beta$</u>. All the convolutions listed above, with the exception of (2.12) and (2.14), are special cases of the so called ψ-products of D. H. Lehmer developed in [24, 26]. Lehmer's important paper [24] on these products, published in 1932, does not seem to have received adequate attention. For this reason, and also because we later introduce a variant of a Lehmer ψ-product, we shall mention it in some detail.

Let $\psi(x,y)$ be a positive integral-valued function defined for a prescribed set T of ordered pairs (x,y), $x,y \in Z$. The ψ-product $\alpha \bigcirc \beta$ of α and β is defined by

(3.1) $$(\alpha \bigcirc \beta)(n) = \sum_{\psi(a,b)=n} \alpha(a)\beta(b) \qquad (n \in Z).$$

Lehmer assumes that ψ satisfies the following postulates.

(3.2) <u>Postulate</u> I. For each $n \in Z$, $\psi(a,b) = n$ has a finite number of solutions.

(3.3) <u>Postulate</u> II. $\psi(a,b) = \psi(b,a)$.

(3.4) <u>Postulate</u> III. $\psi(a,\psi(b,c)) = \psi(\psi(a,b),c)$.

These ensure that $\alpha \bigcirc \beta$ is defined by a finite sum and that $\bigcirc$ is commutative and associative.

(3.5) <u>Postulate</u> IV. For any $n \in Z$, $\psi(a,1) = n \Rightarrow a = n$.

This ensures that the semigroup $(S,\bigcirc)$ has the identity element $\eta(n)$ defined in (2.13).

If $\psi(x,y) = n$ has a solution x for some y, then x is called a ψ-<u>divisor</u> of n, and x and y are called <u>conjugate</u> ψ-<u>divisors</u> of n. Let $d(n)$ denote the largest ψ-divisor of n, and

$$\delta_1(n),\ \delta_2(n),\ \ldots,\ \delta_r(n)$$

be the ψ-conjugates of $d(n)$.

Lehmer assumes another postulate, namely,

(3.6) <u>Postulate</u> V. The equation $d(n) = m$ has for each $m > 0$ one and only one solution $m = n$ and $d(1) = 1$.

He then derives a number of theorems such as the following:

(3.7) The semigroup $(S,\bigcirc)$ has the identity η given by (2.13).

(3.8) α is a unit in $(S,\bigcirc)$ if and only if

$$\sum_{k=1}^{r(n)} \alpha(\delta_k(n)) \neq 0 \qquad (n = 1, 2, \ldots).$$

Lehmer then develops a calculus of ψ-convolution introducing notions such as ψ-multiplicative functions. We shall not go into these details.

Among the examples Lehmer gave of his ψ-products is the product of two functions α, β defined by

$$(\alpha \bigcirc \beta)(1) = \alpha(1)\beta(1),$$

and if $n > 1$ is given by (1.1), then

$$(3.9) \qquad (\alpha \bigcirc \beta)(n) = \sum_{\substack{b_i c_i = a_i + 1 \\ i = 1,\ldots,r}} \alpha\Big(\prod_i p_i^{b_i - 1}\Big) \beta\Big(\prod_i p_i^{c_i - 1}\Big).$$

(3.10) Remark. By varying Lehmer's Postulates I to V on $\psi(x,y)$, it is possible to construct convolutions which vary from Lehmer's products. In particular, the author has replaced Postulate IV by the following:

$$\psi(n,y) = n \Rightarrow y = \gamma(n)$$

where $\gamma(n)$ denotes the core of n with $\gamma(1) = 1$. We shall not go into the details of the new results obtained, but shall consider in detail a special convolution with this property. We call this "exponential convolution" and study it in the next section.

4. Exponential convolution. For arbitrary α and β, we define their exponential product, denoted by $\alpha \odot \beta$, as follows:

$$(\alpha \odot \beta)(1) = \alpha(1)\beta(1);$$

$$(4.1)\qquad (\alpha \odot \beta)(n) = \sum_{\substack{b_j c_j = a_j \\ j = 1,\ldots,r}} \alpha\left(\prod p_j^{b_j}\right) \beta\left(\prod p_j^{c_j}\right), \qquad n > 1,$$

where $n > 1$ has the representation given in (1.1).

(4.2) Remark. This convolution is clearly commutative and associative. This is not of Narkiewicz type, not being of the form $\sum \alpha(d)\beta(n/d)$. In fact, it is not a Lehmer-type product, because it violates his Postulate IV in (3.5).

(4.3) Exponential divisors. If $n = p_1^{a_1} \cdots p_r^{a_r}$, by an exponential divisor of n, we mean a divisor of the form

$$d = p_1^{b_1} \cdots p_r^{b_r}, \qquad (b_j | a_j, \quad j = 1, \ldots, r).$$

We call the divisor $p_1^{a_1/b_1} \cdots p_r^{a_r/b_r}$ the exponential conjugate of d.

We now state some of our results about this convolution.

(4.4) Theorem. (i) The system $(S, \odot)$ is a commutative semigroup with identity element $|\mu|$ (where $|\mu|$ is the arithmetic function defined by $|\mu(n)|$, $\mu(n)$ being the Möbius function).

(ii) The units of $(S, \odot)$ are those α for which $\alpha(n) \neq 0$ whenever n is a product of distinct primes, and $\alpha(1) \neq 0$.

(iii) The semigroup $(S, \odot)$ has an infinity of zero divisors. An element α of $(S, \odot)$ is a non-zero-divisor only if, given any finite number of primes $p_1, \ldots, p_r$, there exist corresponding positive integers $a_1, \ldots, a_r$ such that

$$\alpha\left(p_1^{a_1} \cdots p_r^{a_r}\right) \neq 0.$$

(iv) $(S, \odot)$ has no non-trivial nilpotent elements.

Proof. Result (i) is easily proved. For if $p_1, \ldots, p_r$ are distinct primes, and α any arithmetic function,

$$(\alpha \odot \mu)(p_1 \cdots p_r) = \alpha(p_1 \cdots p_r) |\mu(p_1 \cdots p_r)|$$

$$= \alpha(p_1 \cdots p_r).$$

If $n = p_1^{a_1} \cdots p_r^{a_r}$ and $a_1 \cdots a_r > 1$, then

$$(\alpha \cdot |\mu|)(n) = \sum_{\substack{b_i c_i = a_i \\ i = 1, \ldots, r}} \alpha\left(\prod p_i^{b_i}\right) \left|\mu\left(\prod p_i^{c_i}\right)\right|$$

$$= \alpha(n)\,|\mu(p_1 \cdots p_r)|,$$

since, recalling the definition of $\mu(n)$, the only non-vanishing term in the sum on the right corresponds to the case $c_1 = \cdots = c_r = 1$.

Thus $|\mu|$ is an identity element of $(S, \odot)$, but there cannot be more than one identity.

To prove (ii), first we note that the condition is necessary. For if we denote the inverse of α by α^{-1} whenever it exists, we have for distinct primes $p_1, \ldots, p_r$,

$$1 = |\mu(p_1 \cdots p_r)| = (\alpha \odot \alpha^{-1})(p_1 \cdots p_r)$$

$$= \alpha(p_1 \cdots p_r)\alpha^{-1}(p_1 \cdots p_r),$$

which implies that $\alpha(p_1 \cdots p_r) \neq 0$.

On the other hand, suppose that $\alpha(p_1 \cdots p_r) \neq 0$ for every finite set of primes $p_1, \ldots, p_r$, and $\alpha(1) \neq 0$. We can construct $\alpha^{-1}(n)$ for all n by induction on n. Thus the relation $\alpha(1)\alpha^{-1}(1) = 1$ gives $\alpha^{-1}(1) = 1/\alpha(1)$. Similarly $\alpha^{-1}(2) = 1/\alpha(2)$.

Take any $n = p_1^{a_1} \cdots p_r^{a_r} > 2$ and assume that $\alpha^{-1}(m)$ is known for all $m < n$. Then from the relation

$$|\mu(n)| = (\alpha \odot \alpha^{-1})(n)$$

$$= \alpha(p_1 \cdots p_r)\alpha^{-1}(n) + \sum_{\substack{b_i c_i = a_i \\ b_1 \cdots b_r > 1}} \alpha\left(\prod p_i^{b_i}\right) \alpha^{-1}\left(\prod p_i^{c_i}\right),$$

we can solve for $\alpha^{-1}(n)$ uniquely.

The proof of (iii) is as follows. The stated condition for α to be a non-zero-divisor is necessary. For suppose there exist primes $p_1, \ldots, p_r$ $(r > 0)$ such that

$$\alpha\left(p_1^{a_1}\cdots p_r^{a_r}\right) = 0$$

for all positive integers $a_1, \ldots, a_r$. Define the function β as follows:

$$\beta(p_1\cdots p_r) = 1,$$

$$\beta(n) = 0, \qquad n \neq p_1\cdots p_r.$$

Then

$$(\alpha \odot \beta)(n) = 0 \quad \text{for all} \quad n \in Z,$$

showing that α is a zero divisor.

Remark. The question whether the above condition is also sufficient for a non-zero-divisor remains open.

To prove (iv), we proceed as follows. Suppose there is an α such that

$$\alpha^{(k)} = \alpha \odot \alpha^{(k-1)} \equiv 0.$$

Then we show that

(4.8) $$\alpha(n) \equiv 0.$$

This clearly holds for $n = 1$, and also whenever n is square-free, since then

$$\alpha^{(k)}(n) = (\alpha(n))^k.$$

We next show that (4.8) holds for N_a of the form

$$N_a = p_1^a p_2\cdots p_r \qquad (a > 0).$$

We proceed by induction on a, this being true for $a = 1$. Assume that $\alpha(N_a) \neq 0$ and $\alpha\left(p_1^{a_1}p_2\cdots p_r\right) = 0$ for $a_1 < a$. Then

$$\alpha^{(2)}\left(p_1^{a^2} p_2\cdots p_r\right) = (\alpha\cdot\alpha)\left(p_1^{a^2} p_2\cdots p_r\right)$$

$$= \alpha(N_a)\,\alpha(N_a)$$

$$\neq 0$$

Let b be the smallest positive integer for which

$$\alpha^{(2)}(p_1^b p_2\cdots p_r) \neq 0.$$

Then

$$\alpha^{(3)}\left(p_1^{ab} p_2\cdots p_r\right) = (\alpha \odot \alpha^{(2)})\left(p_1^{ab} p_2\cdots p_r\right)$$

$$= \alpha\left(p_1^a p_2\cdots p_r\right)\,\alpha^{(2)}\left(p_1^b p_2\cdots p_r\right)$$

$$\neq 0.$$

Continuing this argument, we produce an n such that $\alpha^{(k)}(n) \neq 0$, leading to a contradiction. Hence $\alpha(N_a) = 0$ for $a = 1, 2, \ldots$.

We now continue the induction successively on $a_2, a_3, \ldots$ to show that $\alpha\left(p_1^{a_1} p_2^{a_2} p_3\cdots p_r\right) = 0$, etc, and thus for any integer $n = p_1^{a_1}\cdots p_r^{a_r}$, completing the proof.

(4.9) <u>Exponential analogue of Möbius function.</u> Let us define $\mu^{(e)}(n)$, the exponential analogue of the Möbius function, as follows:

$$\mu^{(e)}(1) = 1$$

and for $n > 1$ given by (1.1),

$$\mu^{(e)}(n) = \mu(a_1)\cdots\mu(a_r).$$

Clearly, $\mu^{(e)}(n)$ is a multiplicative function and also exponentially multiplicative (see (1.4)). We can also verify that if $E(n)$ is the function which equals 1 for all n, then its exponential inverse is $\mu^{(e)}(n)$. It follows that for any α and β,

$$\alpha = \beta \odot E \quad \Leftrightarrow \quad \beta = \alpha \odot \mu^{(e)} .$$

We know that in $(S,\cdot)$ the semi-multiplicative functions form a semigroup [30] and the multiplicative functions a group. We note the following analogous result without proof.

(4.10) In $(S,\odot)$ the set of all unit multiplicative functions form a group.

5. The connection between exponential convolution and Dirichlet convolution. The Dirichlet convolution of arithmetic functions $\alpha(n)$ of a single argument can be extended to functions $\alpha(n_1,\dots,n_k)$ of k arguments, k being an arbitrary finite number. In fact such an extension was already considered by Vaidyanathaswamy in [43].

If S_k denotes the set of all arithmetic functions $\alpha(n_1,\dots,n_k)$ of k arguments, we define the Dirichlet product $\alpha\cdot\beta$ by

$$(\alpha\cdot\beta)(n_1,\dots,n_k) = \sum_{\substack{a_j b_j = n_j \\ j = 1,\dots,k}} \alpha(a_1,\dots,a_k)\ \beta(b_1,\dots,b_k).$$

The system $(S_n,\cdot)$ is a commutative semigroup having as the identity the arithmetic function η_k given by

$$\eta_k(n_1,\dots,n_k) = \begin{cases} 1, & n_1 = \cdots = n_k = 1; \\ 0, & \text{otherwise.} \end{cases}$$

The units of $(S_n,\cdot)$ are those functions α for which $\alpha(1,\dots,1) \neq 0$. It can be shown that the system $(S_n,+,\cdot)$ is a commutative ring, and in fact a domain of integrity. Following the method of Cashwell and Everett [8], one can study the unique factorization property for $(S_n,+,\odot)$.

The Möbius function for $(S_n,\cdot)$ is the function $\mu(n_1,\dots,n_k)$ defined by

$$\mu(n_1,\dots,n_k) = \mu(n_1)\cdots\mu(n_k).$$

(See [43, sec. 4].

We wish to point out that some of these results can be extended to the set $\overline{S}$ of complex-valued arithmetic functions $\overline{\alpha} = \overline{\alpha}(\overline{h})$ whose arguments $\overline{h}$ are vectors of the form

$$\overline{h} = (h_1, h_2, \ldots),$$

where $h_1, h_2, \ldots$ are non-negative integers all but a finite number of which are zero. We denote the set of all such vectors by $\overline{Z}$.

Let

$$\overline{s} = (s_1, s_2, \ldots), \tag{5.1}$$

$$\overline{t} = (t_1, t_2, \ldots) \tag{5.2}$$

be two such vectors. We then define that $\overline{s}$ and $\overline{t}$ are of the "same type" if and only if

$$s_j = 0 \Leftrightarrow t_j = 0 \qquad (j = 1, 2, \ldots).$$

If $\overline{s}$ and $\overline{t}$ are of the same type and are given by (5.1) and (5.2), we define the vector $\overline{s}\overline{t}$ as follows:

$$\overline{s}\overline{t} = (s_1 t_1, s_2 t_2, \ldots). \tag{5.3}$$

It should be noted that the vector $\overline{s}\overline{t}$ is not defined unless $\overline{s}$ and $\overline{t}$ are of the same type.

Let $\overline{\alpha}, \overline{\beta}, \in \overline{S}$. We define their Dirichlet product, denoted by $\overline{\alpha}\cdot\overline{\beta}$, by the relation

$$(\overline{\alpha}\cdot\overline{\beta})(\overline{h}) = \sum_{\overline{s}\overline{t}=\overline{h}} \overline{\alpha}(\overline{s})\overline{\beta}(\overline{t}) \qquad (\overline{h} \in Z_o).$$

We recall the definition $\eta(n)$ given in (2.13) and extend it by defining $\eta(0) = 1$. For $\overline{h} = (h_1, \ldots)$, we define the function $\overline{\eta}$ by

$$\overline{\eta}(\overline{h}) = \eta(h_1)\eta(h_2)\cdots . \tag{5.4}$$

Finally, we write $|\overline{h}|$ for $\max_j h_j$, whenever it exists. We now have the following result.

We can now define $\overline{\eta}$ alternately as

(5.5) $$\overline{\eta}(\overline{h}) = \begin{cases} 1, & |\overline{h}| \le 1, \\ 0, & |\overline{h}| > 1. \end{cases}$$

(5.6) <u>Theorem</u>. $(\overline{S},\cdot)$ is a commutative semigroup for which

(i) the identity element is $\overline{\eta}$;

(ii) the units are those $\overline{\alpha}$ for which

$$\overline{\alpha}(\overline{h}) \ne 0 \quad \text{for} \quad |\overline{h}| \le 1;$$

(iii) if $\overline{E} \in \overline{S}$ is defined by $\overline{E}(\overline{h}) = \overline{1}$ for all $\overline{h} \in \overline{Z}$, and if $\overline{\mu}$ is the inverse of $\overline{E}$ (so that $\overline{\mu}(\overline{h})$ is the Möbius function of $(\overline{S},\cdot)$), $\overline{\mu}$ is given by

(5.7) $$\overline{\mu}(\overline{h}) = \mu(h_1)\mu(h_2)\cdots, \qquad \overline{h} = (h_1,h_2,\ldots),$$

with the convention that $\mu(0) = 0$.

The proof is similar to that for the semigroup $(S,\cdot)$, and is omitted.

We next recall that any $n \in Z$ has the representation given in (1.2), so that the mapping

$$n \longleftrightarrow \overline{h} = (h_1,h_2,\ldots)$$

is one-to-one on Z onto $\overline{Z}$. Further, to each $\alpha \in S$, there corresponds in a one-to-one manner the element $\overline{\alpha}$ in $\overline{S}$ given by

(5.8) $$\alpha(n) \longleftrightarrow \overline{\alpha}(\overline{h}), \qquad n = q_1^{h_1}q_2^{h_2}\cdots, \quad \overline{h} = (h_1,h_2,\ldots).$$

Further, this correspondence preserves the semigroup operation, that is, if

$$\alpha \longleftrightarrow \overline{\alpha}, \quad \beta \longleftrightarrow \overline{\beta},$$

then

$$\alpha \odot \beta \longleftrightarrow \overline{\alpha}\cdot\overline{\beta}.$$

Thus we have proved the following result.

(5.9) Theorem. The semigroups $(S, \odot)$ and $(\bar{S}, \cdot)$ are isomorphic to each other.

(5.10) Remark. From the mapping given in (5.8), we could at once deduce the results (i), (ii) of Theorem (4.4) and the results in (4.9) from Theorem (5.6).

6. The order and average order of $\tau^{(e)}(n)$.

We have already defined $\tau^{(e)}(n)$ to be the number of exponential divisors of n, $\tau^{(e)}(1) = 1$.

If $n > 1$ has the representation (1.1), we have

$$\tau^{(e)}(n) = \tau(a_1)\tau(a_2)\cdots\tau(a_r), \tag{6.1}$$

$\tau(a)$ denoting the number of divisors of n. We now prove:

(6.2) Theorem (Erdös).

$$\overline{\lim_{n\to\infty}} \frac{\log \tau^{(e)}(n) \log\log n}{\log n} = \tfrac{1}{2}\log 2.$$

Proof. We first prove that for any given $\epsilon > 0$, there are infinitely many positive integers m for which

$$\tau^{(e)}(m) > 2^{(1-\epsilon)\log m/2\log\log m}.$$

Let $m = q_1^2 q_2^2 \cdots q_k^2$, where $q_1, \ldots, q_k$ are the first k primes. Then by the prime number theorem, $\tfrac{1}{2}\log m \sim y$ where $y = q_k$, and $k = \pi(q_k) = \pi(y) \sim y/\log y$, where $\pi(x)$ = number of primes $\le x$. Hence

$$\begin{aligned}\log \tau^{(e)}(m) &= k \log 2\\ &\sim (\log 2)y/\log y\\ &\sim (\log 2)\tfrac{1}{2}\log m/\log\log m,\end{aligned}$$

from which we get the inequality stated above.

To complete the proof of the theorem, it remains only to show that given any $\epsilon > 0$,

$$\tau^{(e)}(n) < 2^{(1+\epsilon)\log n/2\log\log n} \qquad (n > n_0(\epsilon)).$$

Put

$$F(n) = \max_{1 \le m \le n} \tau^{(e)}(m)$$

and assume that t is the smallest integer for which $\tau^{(e)}(t) = F(n)$.

Put $t = t_1 t_2$ where all the prime factors of t_1 are less than $\log n/(\log\log n)^2$ and all prime factors of t_2 are $\ge \log n/(\log\log n)^2$. We have

$$\tau^{(e)}(t_1) < \left(\frac{\log n}{\log 2}\right)^{\log n/(\log\log n)^3} < 2^{\epsilon \log n/2 \log\log n}$$

for all $n \ge n_0(\epsilon)$, because t_1 has fewer than $(1+\epsilon)\log n/(\log\log n)^3$ prime factors and the exponent of each prime factor in the canonical form of t_1 is $< \log n/\log 2$.

Let us now look at t_2. Put $t_2 = p_1^{b_1} \cdots p_r^{b_r}$, where $p_1, \ldots, p_r$ are consecutive primes $\ge \log n/(\log\log n)^2$. We have

$$b_1 + \cdots + b_r \le (1 + o(1)) \log n/\log\log n$$

since $n \ge t_2 \ge p_1^{b_1+\cdots+b_r}$, or $b_1 + \cdots + b_r \le \log n/\log p_1$. We can assume that all the b_j's are > 1 and are even, since if b is odd, there is an even $c < b$ with $\tau(c) \ge \tau(b)$ (recall the minimal nature of t) and if $b = 1$, it makes no contribution to $\tau(n)$.

Now, if we have even numbers whose sum is given, their product is maximal if all are 2, as can be easily proved, for instance, by induction. Thus

$$\tau^{(e)}(t_2) = \prod_{j=1}^{r} \tau(b_j) \le \prod_{j=1}^{r} b_j$$

$$\le 2^{(1 + o(1))\log n/2 \log\log n},$$

which is what we set out to establish. Theorem (6.2) is thus proved.

Remark. The result of the theorem may be compared with the well known result:

$$\lim_{n\to\infty} \frac{\log \tau(n) \log\log n}{\log n} = \log 2 .$$

(6.3) <u>Theorem</u>. Let

$$T(x) = \sum_{n \le x} \tau^{(e)}(n).$$

Then

$$T(x) = Ax + O(x^{\frac{1}{2}} \log x)$$

where

$$A = \prod_{p} \left\{1 + (1-p^{-1}) \sum_{k=2}^{\infty} (p^k-1)^{-1}\right\}.$$

<u>Proof</u>. Let

$$f(s) = \sum_{n=1}^{\infty} \frac{\tau^{(e)}(n)}{n^s} .$$

This is regular for $\sigma = \mathrm{Re}(s) > 1$. Since $\tau^{(e)}(n)$ is multiplicative, on using (6.1),

$$\begin{aligned} f(s) &= \prod_{p} \left\{1 + \tau(1)p^{-s} + \tau(2)p^{-2s} + \cdots \tau(a)p^{-as} + \cdots\right\} \\ &= \prod_{p} \left\{1 + \frac{p^{-s}}{1-p^{-s}} + \frac{p^{-2s}}{1-p^{-2s}} + \cdots\right\} \\ &= \prod_{p} (1-p^{-s})^{-1}\varphi(s) \\ &= \zeta(s)\varphi(s) \end{aligned}$$

where $\zeta(s)$ is the Riemann zeta function and

$$\varphi(s) = \prod_{p} \left\{1 + (1-p^{-s}) \sum_{k=2}^{\infty} (p^{ks}-1)^{-1}\right\}.$$

Clearly, $\varphi(s)$ is regular for $\sigma > \frac{1}{2}$. An application of Ikehara's theorem now gives

(6.5) $$\lim_{x\to\infty} \left(\frac{T(x)}{x}\right) = \varphi(1) = A.$$

The usual contour integration method, applied to (6.4), gives the order estimate of the error term of the theorem after considerable laborious calculations. The details will be given elsewhere.

(6.6) Remark. We can extend (6.5) as follows:

Let

$$T^{(k)}(x) = \sum_{n \le x} [\tau^{(e)}(n)]^k.$$

Then

$$\lim_{x\to\infty} \left(\frac{T^{(k)}(x)}{x}\right) \text{ exists for every } k \text{ and } = A_k,$$

where

$$A_k = \prod_p \left\{1 + \sum_{n=2}^{\infty} \left[(\tau(n))^k - (\tau(n-1))^k\right] p^{-n}\right\}.$$

Another result of interest is as follows. Let an integer n be said to be exponentially square-free if in its canonical form each exponent is square-free, with the convention that 1 is taken to be exponentially square-free. Let $Q^{(e)}(x)$ denote the number of exponentially square-free integers $\le x$. Then we have

(6.7) Theorem. $$Q^{(e)}(x) = Bx + O(x^{\frac{1}{2}})$$

where

$$B = \prod_p \left(1 - \sum p^{-as} + \sum p^{-bs}\right),$$

where a ranges over all non-square-free numbers for which $a-1$ is square-free, and b ranges over all square-free numbers for which $b-1$ is non-square-free.

<u>Proof</u>. We proceed as in the proof of (6.3) and use a result of H. Delange [16], namely, if $\alpha(n)$ is a multiplicative arithmetic function with $\alpha(n) = 0$ or 1, and for primes p, $\alpha(p) = 1$, then

$$\sum_{n \le x} \alpha(n) = cx + o(x^{\frac{1}{2}}),$$

c being a constant.

7. <u>Exponentially perfect numbers</u>. Many of the usual problems associated with Dirichlet convolution have their counterparts in exponential (or, in fact, any other) convolution. As an example, we mention the question of determination of all exponentially perfect numbers, which by definition are positive integers n for which

$$\sigma^{(e)}(n) = 2n,$$

where $\sigma^{(e)}(n)$ denotes the sum of the exponential divisors of n. An example of such a number is 36. It is easy to see that there exist an infinity of them; for if m is exponentially perfect, so is mk (called the <u>associate</u> of m) where k is any square-free integer relatively prime to m.

Some examples of exponentially perfect numbers are:

$$2^2.3^3.5^2; \quad 2^3.3^2.5^2; \quad 2^4.3^2.11^2;$$
$$2^6.3^2.7^2.13^2; \quad 2^7.3^2.5^2.7^2.13^2;$$
$$2^8.3^2.5^2.7^2.139^2.$$

We raise the following questions.

(7.1) Is there an odd exponentially perfect number?

(7.2) Are there an infinity of exponentially perfect numbers such that no two of them are associates of each other?

We obtained some necessary conditions for an odd integer to be exponentially perfect, which will be published elsewhere.

8. <u>Remark</u>. We finally remark that to every given convolution of arithmetic functions, one can define the corresponding exponential convolution and study the properties of arithmetical functions which arise therefrom. For example, one can study the exponential unitary

convolution, and in fact, the exponential analogue of any Narkiewicz-type convolution, among others.

9. Acknowledgement. The author is thankful to Professor P. Erdős for sending his proof of Theorem (6.2) and to Dr. Suryanarayana for his help in the preparation of the list of references.

REFERENCES

1. E. T. Bell, On Liouville's theorems concerning certain numerical functions, Bull. Amer. Math. Soc. 19 (1912), 164-166.

2. E. T. Bell, Euler Algebra, Trans. Amer. Math. Soc. 25 (1923), 135-154.

3. E. T. Bell, Outline of a theory of arithmetical functions in their algebraic aspects, J. Indian Math. Soc. 17 (1928), 249-260.

4. E. T. Bell, Factorability of numerical functions, Bull. Amer. Math. Soc. 37 (1931), 251-253.

5. L. Carlitz, Rings of arithmetic functions, Pacific J. Math. 14 (1964), 1165-1171.

6. L. Carlitz, Arithmetic functions in an unusual setting, Amer. Math. Monthly 73 (1966), 582-590.

7. L. Carlitz, Arithmetical functions in an unusual setting II, Duke Math. J. 34 (1967), 757-759.

8. E. D. Cashwell and C. J. Everett, The ring of number-theoretic functions, Pacific J. Math. 9 (1959), 975-985.

9. J. Chidambaraswamy, Sum functions of unitary and semi-unitary divisors, J. Indian Math. Soc. 31 (1967), 117-126.

10. E. Cohen, Rings of arithmetic functions, Duke Math. J. 19 (1952), 115-129.

11. E. Cohen, Rings of arithmetic functions II: The number of solutions of quadratic congruences, Duke Math. J. 21 (1954), 9-28.

12. E. Cohen, Unitary products of arithmetical functions, Acta Arith. 7 (1961), 29-38.

13. E. Cohen, Arithmetical functions of finite Abelian groups, Math. Ann. 142 (1961), 165-182.

14. H. H. Crapo, The Möbius function of a lattice, J. Combinatorial Theory 1 (1966), 126-131.

15. T. M. K. Davison, On arithmetic convolutions, Canad. Math. Bull. 9 (1966), 287-296.

16. H. Delange, Sur certaines functions additives á valeur entiers, Acta Arith. 16 (1969/70), 195-206.

17. P. Erdős, (Private communication to the author).

18. G. Gesely, A generalized arithmetic composition, Amer. Math. Monthly 74 (1967), 1216-1217.

19. A. A. Gioia, The k-product of arithmetic functions, Canad. J. Math. 17 (1965), 970-976.

20. A. A. Gioia, On an identity for multiplicative functions, Amer. Math. Monthly 69 (1962), 988-991.

21. A. A. Gioia, Generalized Dirichlet products of number theoretic functions. Ph.D. Thesis, University of Missouri, 1964.

22. D. L. Goldsmith, On the multiplicative properties of arithmetic functions, Pacific J. Math. 27 (1968), 283-304.

23. D. L. Goldsmith, A generalized convolution for arithmetic functions, Duke Math. J. 38 (1971), 279-283.

24. D. H. Lehmer, Arithmetic of double series, Trans. Amer. Math. Soc. 33 (1931), 945-457.

25. D. H. Lehmer, A new calculus of numerical functions, Amer. J. Math. 53 (1931), 843-854.

26. D. H. Lehmer, Polynomials for the n-ary composition of numerical functions, Amer. J. Math. 58 (1936), 563-572.

27. P. J. McCarthy, Regular arithmetical convolutions, Portugal. Math. 10 (1963), 81-94.

28. W. Narkiewicz, On a class of arithmetical convolutions, Colloq. Math. 10 (1963), 81-94.

29. J. Popken, On convolutions in number theory, Indag. Math. 17 (1955), 10-15.

30. D. Rearick, Semi multiplicative functions, Duke Math. J. 33 (1966), 49-54.

31. D. Rearick, Operations on algebras of arithmetic functions, Duke Math. J. 35 (1968), 761-766.

32. G. C. Rota, On the foundations of combinatorial theory I. Theory of Möbius functions, Z. Wahrscheinlichkeitstheorie und Verw. 2 (1964), 340-368.

33. R. Siva Ramakrishna, Contributions to the study of multiplicative arithmetic functions, (to appear).

34. V. Sivaramprasad and D. Suryanarayana, Sum functions of k-ary and semi-k-ary divisors, J. Austral. Math. Soc. (to appear).

35. D. A. Smith, Incidence functions as generalized arithmetic functions I, Duke Math. J. 34 (1967), 617-633.

36. D. A. Smith, Incidence functions as generalized arithmetic functions II, Duke Math. J. 36 (1969), 15-30.

37. D. A. Smith, Incidence functions as generalized arithmetic functions III, Duke Math. J. 36 (1969), 353-367.

38. M. V. Subbarao and A. A. Gioia, Generalized Dirichlet products of arithmetical functions (abstract), Notices Amer. Math. Soc. 10 (1963), No. 7, p. 661.

39. M. V. Subbarao and A. A. Gioia, Identities for multiplicative functions, Canad. Math. Bull. 10 (1967), 65-73.

40. M. Sugunamma, Contributions to the study of general arithmetic functions, Ph.D. Thesis, Sri Venkateswara University, Tirupati, India, 1965.

41. D. Suryanarayana, The number of bi-unitary divisors of an integer, Proceedings of a Conference on the Theory of Arithmetic Functions, Kalamazoo, 1971.

42. R. Vaidyanathaswamy, The identical equations of the multiplicative function, Bull. Amer. Math. Soc. 36 (1930), 762-772.

43. R. Vaidyanathaswamy, The theory of multiplicative arithmetical functions, Trans. Amer. Math. Soc. 33 (1931), 579-662.

44. L. Weisner, Abstract theory of inversion of finite series, Trans. Amer. Math. Soc. 38 (1935), 474-484.

THE NUMBER OF BI-UNITARY DIVISORS OF AN INTEGER

D. Suryanarayana*, University of Alberta

1. Introduction. It is well-known that a divisor $d > 0$ of the positive integer n is called *unitary* if $d\delta = n$ and $(d, \delta) = 1$. We write $d \| n$ to mean that d is a unitary divisor of n. For integers a, b not both zero, let us define $(a,b)^{**}$ to be the greatest unitary divisor of both a and b. In 1960, E. Cohen [2] defined $(a,b)^{*}$ to be the greatest divisor of a which is a unitary divisor of b. It is clear that $(a,b)^{**} = (b,a)^{**}$ for all a and b, but $(a,b)^{*} \neq (b,a)^{*}$ for some a and b. In 1967, J. Chidambaraswamy [1] defined a divisor $d > 0$ of the positive integer n to be *semi-unitary* if $d\delta = n$ and $(d, \delta)^{*} = 1$. We call a divisor $d > 0$ of the positive integer n *bi-unitary* if $d\delta = n$ and $(d, \delta)^{**} = 1$. It is clear that the conjugate divisor of a bi-unitary divisor of n is also bi-unitary, whereas the conjugate divisor of a semi-unitary divisor of n need not be semi-unitary. Let $\tau^{**}(n)$ denote the number of bi-unitary divisors of n.

The object of this paper is to prove the following

Theorem. For $x \geq 2$,

$$\sum_{n \leq x} \tau^{**}(n) = Ax\left(\log x + 2\gamma - 1 + 2 \sum_{p} \frac{(p-1)^2 p^2 \log p}{(p^2+1)(p^4+2p-1)} + 2B\right)$$

$$+ O(x^{\frac{1}{2}} \log x),$$

where γ *is Euler's constant and the constants* A *and* B *are given by* (2.6) *and* (2.11).

2. Prerequisites. In this section, we prove some lemmas which are needed in the proof of the theorem.

---

* On leave from Andhra University, Waltair, India

Lemma 2.1. $\tau^{**}(n)$ is a multiplicative function of n.

Proof. It is clear that $\tau^{**}(1) = 1$. Let $(m,n) = 1$ and let $d|mn$. We can write $d = d_1 d_2$, where $d_1|m$, $d_2|n$ and $(d_1,d_2) = 1$. We shall now prove the following:

$$(2.1) \qquad (d,\frac{mn}{d})^{**} = (d_1,\frac{m}{d_1})^{**}(d_2,\frac{n}{d_2})^{**}.$$

Let $r = (d_1,\frac{m}{d_1})^{**}$, $s = (d_2,\frac{n}{d_2})^{**}$ and $t = (d,\frac{mn}{d})^{**}$. Then $r\|d_1$, $r\|\frac{m}{d_1}$; $s\|d_2$, $s\|\frac{n}{d_2}$. Since $(d_1,d_2) = 1$ and $(\frac{m}{d_1},\frac{m}{d_2}) = 1$, we have $rs\|d_1d_2$, $rs\|\frac{mn}{d_1d_2}$, so that $rs|(d_1d_2,\frac{mn}{d_1d_2})$, that is, $rs|t$. Also, since $t = (d,\frac{mn}{d})^{**}$, we have $t\|d_1d_2$, $t\|\frac{mn}{d_1d_2}$. Since $(d_1,d_2) = 1$ and $t\|d_1d_2$, we can write $t = t_1t_2$, where $t_1\|d_1$, $t_2\|d_2$. Since $t\|\frac{mn}{d_1d_2}$ and $(\frac{m}{d_1},\frac{n}{d_2}) = 1$, we have $t_1\|\frac{m}{d_1}$, $t_2\|\frac{n}{d_2}$. Hence $t_1|(d_1,\frac{m}{d_1})^{**}$ and $t_2|(d_2,\frac{n}{d_2})^{**}$, so that $t_1t_2|rs$, that is, $t|rs$.

Now since $rs|t$ and $t|rs$, we have $t = rs$, so that (2.1) follows.

From (2.1), we have $(d,\frac{mn}{d})^{**} = 1$ if and only if $(d_1,\frac{m}{d_1})^{**} = 1$ and $(d_2,\frac{n}{d_2})^{**} = 1$. Hence $\tau^{**}(mn) = \tau^{**}(m)\tau^{**}(n)$, so that Lemma 2.1 follows.

Lemma 2.2. If $n = \prod_{i=1}^{r} p_i^{\alpha_i}$ is the canonical representation of $n > 1$, then

$$\tau^{**}(n) = \Big\{\prod_{\alpha_i \text{ is even}} (\alpha_i)\Big\}\Big\{\prod_{\alpha_i \text{ is odd}} (1+\alpha_i)\Big\}.$$

Proof. In virtue of Lemma 2.1, it is enough, if we prove that $\tau^{**}(p^\alpha) = \alpha$ or $1+\alpha$ according as α is even or α is odd. Now, if α is odd, for every divisor $d = p^t$ of p^α, we have $(p^t,p^{\alpha-t})^{**} = 1$, so that $\tau^{**}(p^\alpha) = 1+\alpha$. If α is even, $\alpha = 2\beta$, say, then for every divisor $d = p^t$ of p^α, we have $(p^t,p^{\alpha-t})^{**} = 1$, except for $t = \beta$, so that $\tau^{**}(p^\alpha) = (1+\alpha) - 1 = \alpha$.

Hence Lemma 2.2 follows.

Lemma 2.3 (cf. [6], Theorem 2.4, $k = 2$). *If* $f(n)$ *and* $g(n)$ *are multiplicative, then*

$$h(n) = \sum_{\substack{d^2\delta=n \\ (d,\delta)=1}} f(d)g(\delta),$$

is also multiplicative.

Lemma 2.4.
$$\tau^{**}(n) = \sum_{\substack{d^2\delta=n \\ (d,\delta)=1}} \mu^*(d)\tau(\delta),$$

where $\mu^*(n)$ *is the unitary analogue of the Möbius* μ*-function defined by* $\mu^*(n) = (-1)^{\omega(n)}$, $\omega(n)$ *being the number of distinct prime factors of* n, $\omega(1) = 0$ *and* $\tau(n)$ *is the number of divisors of* n.

Proof. Since $\mu^*(n)$ and $\tau(n)$ are multiplicative, it follows by Lemma 2.3 that the right hand side sum of the lemma is a multiplicative function of n. By Lemma 2.1, $\tau^{**}(n)$ is a multiplicative function of n. Hence, it is enough if we verify the identity for $n = p^\alpha$, a prime power.

If α is even,

$$\sum_{\substack{d^2\delta=p^\alpha \\ (d,\delta)=1}} \mu^*(d)\tau(\delta) = \mu^*(1)\tau(p^\alpha) + \mu^*(p^{\alpha/2})\tau(1)$$

$$= 1 + \alpha - 1 = \alpha.$$

If α is odd,

$$\sum_{\substack{d^2\delta=p^\alpha \\ (d,\delta)=1}} \mu^*(d)\tau(\delta) = \mu^*(1)\tau(p^\alpha) = 1 + \alpha.$$

Since by Lemma 2.2, $\tau^{**}(p^\alpha) = \alpha$ or $\alpha + 1$, according as α is even or α is odd, we have

$$\tau^{**}(p^{\alpha}) = \sum_{\substack{d^2\delta=p^{\alpha} \\ (d,\delta)=1}} \mu^*(d)\tau(\delta).$$

Hence Lemma 2.4 follows.

Throughout the following, x denotes a real variable ≥ 2 and γ denotes Euler's constant.

<u>Lemma</u> 2.5 (Dirichlet divisor problem).

$$\sum_{n \leq x} \tau(n) = x(\log x + 2\gamma - 1) + O(x^{\alpha}). \tag{2.2}$$

It is known (cf. [4], p. 272) that $1/4 < \alpha < 1/3$. The best known result was obtained recently by G. A. Kolesnik [5], who proved that the error term in (2.2) is $O(x^{12/37+\varepsilon})$, where $\varepsilon > 0$. There is a conjecture that $\alpha = 1/4 + \varepsilon$. However, in this paper we need only the fact that $\alpha < 1/2$.

<u>Lemma</u> 2.6 (cf. [3], Lemma 3).

$$\sum_{\substack{m \leq x \\ (m,n)=1}} \tau(m) = \left(\frac{\varphi(n)}{n}\right)^2 \{\log x + 2\gamma - 1 + \alpha(n)\} + O\Big(\sum_{d|n} \frac{3^{\omega(d)}}{d^{\alpha}} x^{\alpha}\Big), \tag{2.3}$$

<u>where $\varphi(n)$ is the Euler totient function,</u> α <u>is the number which appears in the Dirichlet divisor problem and</u> $\alpha(n) = \sum_{p|n} \frac{\log p}{p-1}$.

<u>Remark</u>. B. Gordon and K. Rogers [3] proved the asymptotic formula (2.3) for square-free n with an error term $O\Big(\sum_{d|n} 3^{\omega(d)}\sqrt{x/d}\Big)$. However, using the same argument adopted by them in the proof of their Lemma 3, we can get (2.3) for any n, not necessarily square-free. We only have to use their Lemma 2 with error term replaced by $O\big(3^{\omega(d)}(x/d)^{\alpha}\big)$. In fact, they have pointed out that this replacement could be done. Since this replacement did not serve their purpose, they used (2.2) above with error term $O(x^{\frac{1}{2}})$.

<u>Lemma</u> 2.7. <u>For</u> $s > 3$,

$$\sum_{n=1}^{\infty} \frac{\mu^*(n)\varphi^2(n)}{n^s} = \prod_p \left\{1 - \frac{(p-1)^2}{p^s + p^2}\right\}, \tag{2.4}$$

<u>where the product on the right hand side is extended over all primes</u> p.

<u>Proof</u>. Since $\varphi(n) \le n$ and $|\mu^*(n)| = 1$, the series is absolutely convergent for $s > 3$. Further, the general term of the series is multiplicative. Hence the series can be expanded into an infinite product of Euler type (cf. [4], Theorem 286), so that

$$\sum_{n=1}^{\infty} \frac{\mu^*(n)\varphi^2(n)}{n^s} = \prod_p \left\{\sum_{i=0}^{\infty} \frac{\mu^*(p^i)\varphi^2(p^i)}{p^{is}}\right\}$$

$$= \prod_p \left\{1 - \frac{p^2\left(1 - \frac{1}{p}\right)^2}{p^s} + \frac{p^4\left(1 - \frac{1}{p}\right)^2}{p^{2s}} - \cdots\right\}$$

$$= \prod_p \left\{1 - \frac{\left(1 - \frac{1}{p}\right)^2}{p^{s-2}}\left[1 - \frac{1}{p^{s-2}} + \frac{1}{p^{2s-4}} - \cdots\right]\right\}$$

$$= \prod_p \left\{1 - \frac{(p-1)^2}{p^s} \cdot \frac{1}{1 + p^{-s+2}}\right\}$$

$$= \prod_p \left\{1 - \frac{(p-1)^2}{p^s + p^2}\right\}.$$

Hence Lemma 2.7 follows.

<u>Lemma</u> 2.8. <u>For</u> $s > 3$,

$$\sum_{n=1}^{\infty} \frac{\mu^*(n)\varphi^2(n)\log n}{n^s} = -\prod_p \left\{1 - \frac{(p-1)^2}{p^s + p^2}\right\} \sum_p \frac{(p-1)^2\, p^{s-2}\log p}{(p^{s-2}+1)(p^s+2p-1)}. \tag{2.5}$$

<u>Proof</u>. This series is uniformly convergent for $s \ge 3 + \varepsilon > 3$ and so by termwise differentiation of the series in (2.4) with

respect to s, we get (2.5). For finding the derivative of the right hand side expression of (2.4) with respect to s, we write

$$f(s) = \prod_p \left\{1 - \frac{(p-1)^2}{p^s + p^2}\right\}.$$

Then

$$\log f(s) = \sum_p \log \left\{1 - \frac{(p-1)^2}{p^s + p^2}\right\},$$

so that

$$\frac{f'(s)}{f(s)} = \sum_p \frac{(p-1)^2 \, p^{s-2} \log p}{(p^{s-2} + 1)(p^s + 2p - 1)},$$

and this gives us

$$f'(s) = \prod_p \left\{1 - \frac{(p-1)^2}{p^s + p^2}\right\} \sum_p \frac{(p-1)^2 \, p^{s-2} \log p}{(p^{s-2} + 1)(p^s + 2p - 1)}.$$

Hence Lemma 2.8 follows.

As particular cases of Lemmas 2.7 and 2.8 for $s = 4$, we have the following:

$$\text{(2.6)} \qquad \sum_{n=1}^{\infty} \frac{\mu^*(n)\varphi^2(n)}{n^4} = \prod_p \left\{1 - \frac{(p-1)^2}{p^2(p^2+1)}\right\} = A,$$

say.

$$\text{(2.7)} \qquad \sum_{n=1}^{\infty} \frac{\mu^*(n)\varphi^2(n) \log n}{n^4} = -A \sum_p \frac{(p-1)^2 \, p^2 \log p}{(p^2+1)(p^4+2p-1)}.$$

Lemma 2.9.

(2.8)
$$\sum_{n \le x} \frac{\mu^*(n)\varphi^2(n)}{n^4} = A + O\left(\frac{1}{x}\right).$$

(2.9)
$$\sum_{n \le x} \frac{\mu^*(n)\varphi^2(n)\log n}{n^4} = -A\sum_{p} \frac{(p-1)^2 p^2 \log p}{(p^2+1)(p^4+2p-1)} + O\left(\frac{\log x}{x}\right),$$

where A is the constant given by (2.6).

Proof. Since $\varphi(n) \le n$, we have

$$\sum_{n > x} \frac{\mu^*(n)\varphi^2(n)}{n^4} = O\left(\sum_{n > x} \frac{1}{n^2}\right) = O\left(\frac{1}{x}\right),$$

$$\sum_{n > x} \frac{\mu^*(n)\varphi^2(n)\log n}{n^4} = O\left(\sum_{n > x} \frac{\log n}{n^2}\right) = O\left(\frac{\log x}{x}\right).$$

Hence (2.8) and (2.9) follow by (2.6) and (2.7).

Lemma 2.10. The series $\sum_{n=1}^{\infty} \frac{\mu^*(n)\varphi^2(n)\alpha(n)}{n^4}$ is absolutely convergent.

Proof. We have $\alpha(n) = \sum_{p|n} \frac{\log p}{p-1} < \sum_{p|n} 1 < \tau(n)$. Since $\varphi(n) \le n$, it follows that

$$\left|\frac{\mu^*(n)\varphi^2(n)\alpha(n)}{n^4}\right| \le \frac{\tau(n)}{n^2}.$$

Since $\sum_{n=1}^{\infty} \frac{\tau(n)}{n^2}$ is convergent, Lemma 2.10 follows.

Lemma 2.11.

(2.10)
$$\sum_{n \le x} \frac{\mu^*(n)\varphi^2(n)\alpha(n)}{n^4} = AB + O\left(\frac{\log x}{x}\right),$$

where A is the constant defined by (2.6), and

$$(2.11) \qquad B = \frac{1}{A} \sum_{n=1}^{\infty} \frac{\mu^*(n)\varphi^2(n)\alpha(n)}{n^4}.$$

<u>Proof.</u> We have $\alpha(n) < \tau(n)$ and $\varphi(n) \le n$, so that

$$\sum_{n > x} \frac{\mu^*(n)\varphi^2(n)\alpha(n)}{n^4} = O\Big(\sum_{n > x} \frac{\tau(n)}{n^2}\Big) = O\Big(\frac{\log x}{x}\Big),$$

by partial summation, since $\sum_{n \le x} \tau(n) = O(x \log x)$. Hence Lemma 2.11 follows.

3. <u>Proof of the theorem.</u> By Lemma 2.4, we have

$$\sum_{n \le x} \tau^{**}(n) = \sum_{n \le x} \sum_{\substack{d^2\delta = n \\ (d,\delta)=1}} \mu^*(d)\tau(\delta) + \sum_{\substack{d^2\delta \le x \\ (d,\delta)=1}} \mu^*(d)\tau(\delta)$$

$$= \sum_{d \le \sqrt{x}} \mu^*(d) \sum_{\substack{\delta \le \frac{x}{d^2} \\ (\delta,d)=1}} \tau(\delta).$$

We use the asymptotic formula (2.3) with error term replaced by $O(\tau(n)x^{\alpha})$, which is sufficient for our purpose. Clearly,

$$\sum_{d|n} \frac{3^{\omega(d)}}{d^{\alpha}} = \prod_{p|n} \Big(1 + \frac{3}{p^{\alpha}}\Big) < C \prod_{p|n} \{2\} = C2^{\omega(n)} \le C\tau(n),$$

where C is a positive constant. Hence by (2.3),

$$\sum_{n \le x} \tau^{**}(n) = \sum_{d \le \sqrt{x}} \mu^*(d) \Big\{\frac{x}{d^2}\Big(\frac{\varphi(d)}{d}\Big)^2\Big(\log \frac{x}{d^2} + 2\gamma - 1 + 2\alpha(d)\Big) + O\Big(\tau(d)\frac{x^{\alpha}}{d^{2\alpha}}\Big)\Big\}$$

$$= x(\log x + 2\gamma - 1) \sum_{n \le \sqrt{x}} \frac{\mu^*(n)\varphi^2(n)}{n^4}$$

$$- 2x \sum_{n \le \sqrt{x}} \frac{\mu^*(n)\varphi^2(n)\log n}{n^4}$$

$$+ 2x \sum_{n \le \sqrt{x}} \frac{\mu^*(n)\varphi^2(n)\alpha(n)}{n^4} + O\Big(x^{\alpha} \sum_{n \le \sqrt{x}} \frac{\tau(n)}{n^{2\alpha}}\Big).$$

Now, by (2.8), (2.9), (2.10), we have

$$\sum_{n \le x} \tau^{**}(n) = x(\log x + 2\gamma - 1)\Big\{A + O\Big(\frac{1}{\sqrt{x}}\Big)\Big\}$$

$$- 2x\Big\{-A \sum_{p} \frac{(p-1)^2 p^2 \log p}{(p^2+1)(p^4+2p-1)} + O\Big(\frac{\log x}{\sqrt{x}}\Big)\Big\}$$

$$+ 2x\Big\{AB + O\Big(\frac{\log x}{\sqrt{x}}\Big)\Big\} + O\Big(x^{\alpha} \sum_{n \le \sqrt{x}} \frac{\tau(n)}{n^{2\alpha}}\Big)$$

(3.1) $$= Ax\Big(\log x + 2\gamma - 1 + 2\sum_{p} \frac{(p-1)^2 p^2 \log p}{(p^2+1)(p^4+2p-1)}\Big)$$

$$+ O(x^{1/2}\log x) + O\Big(x^{\alpha} \sum_{n \le \sqrt{x}} \frac{\tau(n)}{n^{2\alpha}}\Big).$$

Now, since $2\alpha < 1$, we have

$$\sum_{n \le \sqrt{x}} \frac{\tau(n)}{n^{2\alpha}} = \sum_{n \le \sqrt{x}} \sum_{d\delta = n} \frac{1}{d^{2\alpha}\delta^{2\alpha}} = \sum_{d\delta \le \sqrt{x}} \frac{1}{d^{2\alpha}\delta^{2\alpha}}$$

$$= \sum_{d \le \sqrt{x}} \frac{1}{d^{2\alpha}} \sum_{\delta \le \frac{\sqrt{x}}{d}} \delta^{-2\alpha} = O\Big(\sum_{d \le \sqrt{x}} \frac{1}{d^{2\alpha}}\Big(\frac{\sqrt{x}}{d}\Big)^{1-2\alpha}\Big)$$

$$= O\Big(x^{\frac{1}{2}-\alpha} \sum_{d \le \sqrt{x}} \frac{1}{d}\Big) = O\big(x^{\frac{1}{2}-\alpha} \log x\big),$$

so that the last O-term in (3.1) above is also $O(x^{\frac{1}{2}} \log x)$.

Thus, the proof of the theorem is complete.

REFERENCES

1. J. Chidambaraswamy, Sum functions of unitary and semi-unitary divisors, J. Indian Math. Soc. 31 (1967), 117-126.

2. E. Cohen, Arithmetical functions associated with the unitary divisors of an integer, Math. Z. 72 (1960), 66-80.

3. B. Gordon and K. Rogers, Sums of the divisor function, Canad. J. Math. 16 (1964), 151-158.

4. G. H. Hardy and E. M. Wright, An Introduction to the Theory of Numbers, Fourth edition, The Clarendon Press, Oxford, 1965.

5. G. A. Kolesnik, An improvement of the remainder term in the divisors problem, Mat. Zametki 6 (1969), 545-554 = Math. Notes of Sciences of the USSR 6 (1969), 784-791.

6. D. Suryanarayana, Two arithmetic functions and asymptotic densities of related sets, Portugal. Math. (to appear).

DENSITY BOUNDS FOR THE SUM OF DIVISORS FUNCTION

Charles R. Wall, East Texas State University

1. Introduction. Let σ be the sum of divisors function, and let φ be Euler's function.

If S is a subset of the positive integers, we define $S(N)$ to be the number of elements of S among the first N positive integers. The density δS of S is defined by

$$\delta S = \lim_{N\to\infty} S(N)/N,$$

provided the limit exists.

In this paper we present upper and lower bounds for the function

$$A(x) = \delta\{n: \sigma(n)/n \geq x\}$$

of a real variable x. Davenport [2] proved that $A(x)$ exists, and is continuous, for all real x. Our technique is a refinement of that used by Behrend [1] to bound $A(2)$, the density of the abundant numbers. Our calculations were performed on an IBM 360/40 computer, and required 26 minutes of machine time.

2. Techniques. For j and k positive integers we define

$$A(x,j,k) = \delta\{n: j|n, \quad (n/j,k) = 1 \quad \text{and} \quad \sigma(n)/n \geq x\}.$$

The existence of the functions $A(x,j,k)$ may be proved by a slight modification of Davenport's proof [2] of the existence of

$$A(x) = A(x,1,1).$$

We may, without loss of generality, take k to be squarefree here. It is not difficult to show that if k is a fixed squarefree integer, then

$$A(x) = \sum A(x,j,k), \tag{1}$$

where the summation is over all j composed of the primes which divide k. The same proof shows that if k divides k', then

(2) $$A(x,1,k) = \sum A(x,j,k'),$$

where the summation is over all j composed of primes dividing k' but not k.

Assume henceforth that k is squarefree and that any prime divisor of j also divides k. Behrend [1] showed that if $\sigma(j)/j \geq x$, then

(3) $$A(x,j,k) = \varphi(k)/jk \qquad (x \leq \sigma(j)/j)$$

and if $\sigma(j)/j < x$, then

(4) $$A(x,j,k) \leq \frac{\sigma(j)/j}{x - \sigma(j)/j} \cdot \frac{M}{j} \qquad (x > \sigma(j)/j)$$

where

$$M = \frac{\varphi(k)}{k} \left\{-1 + \frac{\pi^2}{6} \prod_{p|k} \left(1 - p^{-2}\right)\right\}.$$

We also have the trivial bound

$$A(x,j,k) \leq \varphi(k)/jk,$$

which is better than (4) if $x - \sigma(j)/j$ is small and positive.

Dedekind's function ψ is defined by

$$\psi(n) = n \prod_{p|n} (1 + 1/p);$$

this function is clearly multiplicative, whence

(5) $$\psi(n) \leq \sigma(n)$$

with equality if and only if n is squarefree. It can be shown with little difficulty that

(6) $$2\psi(n) \geq n + \sigma(n).$$

Wall [3] has investigated the functions

$$B(x,j,k) = \delta\{n\colon j|n, \quad (n/j,k) = 1 \text{ and } \psi(n)/n \geq x\},$$

all of which exist and are continuous, and $B(x) = B(x,1,1)$. It is clear from (5) and (6) that

$$B(x,j,k) \le A(x,j,k) \le B\left(\frac{x+1}{2},j,k\right) \tag{7}$$

for all x, j and k. The relationship (7) is significant because the functions $B(x,j,k)$ are easier to bound than are the $A(x,j,k)$.

We set $k = 2\cdot 3\cdot 5\cdot 7\cdot 11\cdot 13\cdot 17$ and let j be composed of the primes 7, 11, 13 and 17 with the exponent on each prime dividing j being 0 or 1 or at least 2. Using (2), (3) and (4) we obtained upper and lower bounds for $A(x,1,30)$. We sharpened these bounds by use of (7) whenever possible.

Suppose that every prime which divides j also divides k, and define y by $y\sigma(j)/j = x$. It is not difficult to show that

$$A(x,j,k) = j^{-1}A(y,1,k). \tag{8}$$

Using our bounds for $A(x,1,30)$ with (1) and (8) and $k = 30$, we obtained our bounds for $A(x)$ by summing over

$$j = 2^a 3^b 5^c,$$

subject to $0 \le a \le 6$ or $a \ge 7$, $0 \le b \le 3$ or $b \ge 4$, and $0 \le c \le 2$ or $c \ge 3$.

3. Results. Our bounds are illustrated by Figure 1; the numerical bounds will be presented elsewhere. Of particular interest is the density of the abundant numbers: we found $0.2441 < A(2) < 0.2909$, a slight improvement of the bounds 0.241 and 0.314 given by Behrend [1].

The author wishes to express his gratitude to research assistants Philip L. Crews and Donald B. Johnson, both undergraduates at East Texas State University.

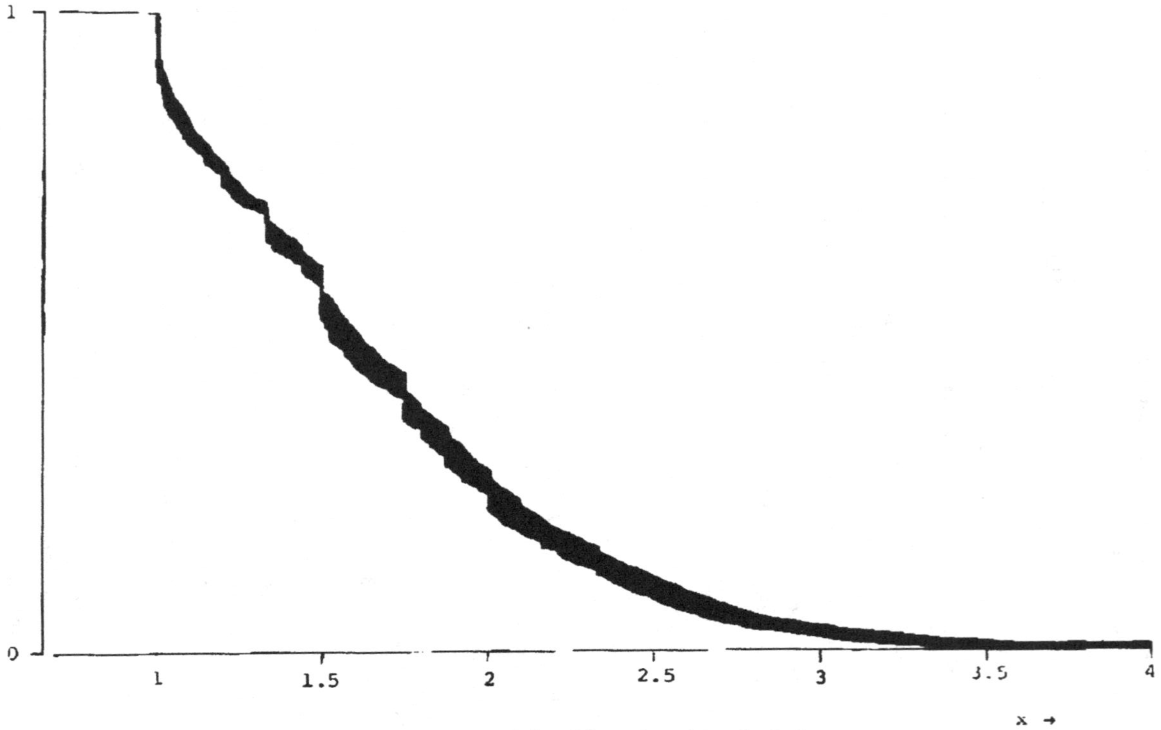

FIGURE 1. A(x) lies in the shaded area.

REFERENCES

1. F. Behrend, Über numeri abundantes, II, Preuss. Akad. Wiss. Sitzungsber. 6 (1933), 280-293

2. H. Davenport, Über numeri abundantes, Preuss. Akad. Wiss. Sitzungsber. 26/29 (1933), 830-837.

3. C. Wall, Topics Related to the Sum of Unitary Divisors of an Integer, Ph.D. Dissertation, Univ. of Tennessee, Knoxville, 1970.

Lecture Notes in Mathematics

Comprehensive leaflet on request

Vol. 38: R. Berger, R. Kiehl, E. Kunz und H.-J. Nastold, Differentialrechnung in der analytischen Geometrie IV, 134 Seiten. 1967. DM 16,-

Vol. 39: Séminaire de Probabilités I. II, 189 pages. 1967. DM 16,-

Vol. 40: J. Tits, Tabellen zu den einfachen Lie Gruppen und ihren Darstellungen. VI, 53 Seiten. 1967. DM 16,-

Vol. 41: A. Grothendieck, Local Cohomology. VI, 106 pages. 1967. DM 16,-

Vol. 42: J. F. Berglund and K. H. Hofmann, Compact Semitopological Semigroups and Weakly Almost Periodic Functions. VI, 160 pages. 1967. DM 16,-

Vol. 43: D. G. Quillen, Homotopical Algebra VI, 157 pages. 1967. DM 16,-

Vol. 44: K. Urbanik, Lectures on Prediction Theory. IV, 50 pages. 1967. DM 16,-

Vol. 45: A. Wilansky, Topics in Functional Analysis. VI, 102 pages. 1967. DM 16,-

Vol. 46: P. E. Conner, Seminar on Periodic Maps. IV, 116 pages. 1967. DM 16,-

Vol. 47: Reports of the Midwest Category Seminar I. IV, 181 pages. 1967. DM 16,-

Vol. 48: G. de Rham, S. Maumary et M. A. Kervaire, Torsion et Type Simple d'Homotopie. IV, 101 pages. 1967. DM 16,-

Vol. 49: C. Faith, Lectures on Injective Modules and Quotient Rings. XVI, 140 pages. 1967. DM 16,-

Vol. 50: L. Zalcman, Analytic Capacity and Rational Approximation. VI, 155 pages. 1968. DM 16,-

Vol. 51: Séminaire de Probabilités II. IV, 199 pages. 1968. DM 16,-

Vol. 52: D. J. Simms, Lie Groups and Quantum Mechanics. IV, 90 pages. 1968. DM 16,-

Vol. 53: J. Cerf, Sur les difféomorphismes de la sphère de dimension trois (Γ_4 = O). XII, 133 pages. 1968. DM 16,-

Vol. 54: G. Shimura, Automorphic Functions and Number Theory. VI, 69 pages. 1968. DM 16,-

Vol. 55: D. Gromoll, W. Klingenberg und W. Meyer, Riemannsche Geometrie im Großen. VI, 287 Seiten. 1968. DM 20,-

Vol. 56: K. Floret und J. Wloka, Einführung in die Theorie der lokalkonvexen Räume. VIII, 194 Seiten. 1968. DM 16,-

Vol. 57: F. Hirzebruch und K. H. Mayer, O (n)-Mannigfaltigkeiten, exotische Sphären und Singularitäten. IV, 132 Seiten. 1968. DM 16,-

Vol. 58: Kuramochi Boundaries of Riemann Surfaces. IV, 102 pages. 1968. DM 16,-

Vol. 59: K. Jänich, Differenzierbare G-Mannigfaltigkeiten. VI, 89 Seiten. 1968. DM 16,-

Vol. 60: Seminar on Differential Equations and Dynamical Systems. Edited by G. S. Jones. VI, 106 pages. 1968. DM 16,-

Vol. 61: Reports of the Midwest Category Seminar II. IV, 91 pages. 1968. DM 16,-

Vol. 62: Harish-Chandra, Automorphic Forms on Semisimple Lie Groups X, 138 pages. 1968. DM 16,-

Vol. 63: F. Albrecht, Topics in Control Theory. IV, 65 pages. 1968. DM 16,-

Vol. 64: H. Berens, Interpolationsmethoden zur Behandlung von Approximationsprozessen auf Banachräumen. VI, 90 Seiten. 1968. DM 16,-

Vol. 65: D. Kölzow, Differentiation von Maßen. XII, 102 Seiten. 1968. DM 16,-

Vol. 66: D. Ferus, Totale Absolutkrümmung in Differentialgeometrie und -topologie. VI, 85 Seiten. 1968. DM 16,-

Vol. 67: F. Kamber and P. Tondeur, Flat Manifolds. IV, 53 pages. 1968. DM 16,-

Vol. 68: N. Boboc et P. Mustată, Espaces harmoniques associés aux opérateurs différentiels linéaires du second ordre de type elliptique. VI, 95 pages. 1968. DM 16,-

Vol. 69: Seminar über Potentialtheorie. Herausgegeben von H. Bauer. VI, 180 Seiten. 1968. DM 16,-

Vol. 70: Proceedings of the Summer School in Logic. Edited by M. H. Löb. IV, 331 pages. 1968. DM 20,-

Vol. 71: Séminaire Pierre Lelong (Analyse), Année 1967 - 1968. VI, 190 pages. 1968. DM 16,-

Vol. 72: The Syntax and Semantics of Infinitary Languages. Edited by J. Barwise. IV, 268 pages. 1968. DM 18,-

Vol. 73: P. E. Conner, Lectures on the Action of a Finite Group. IV, 123 pages. 1968. DM 16,-

Vol. 74: A. Fröhlich, Formal Groups. IV, 140 pages. 1968. DM 16,-

Vol. 75: G. Lumer, Algèbres de fonctions et espaces de Hardy. VI, 80 pages. 1968. DM 16,-

Vol. 76: R. G. Swan, Algebraic K-Theory. IV, 262 pages. 1968. DM 18,-

Vol. 77: P.-A. Meyer, Processus de Markov: la frontière de Martin. IV, 123 pages. 1968. DM 16,-

Vol. 78: H. Herrlich, Topologische Reflexionen und Coreflexionen. XVI, 166 Seiten. 1968. DM 16,-

Vol. 79: A. Grothendieck, Catégories Cofibrées Additives et Complexe Cotangent Relatif. IV, 167 pages. 1968. DM 16,-

Vol. 80: Seminar on Triples and Categorical Homology Theory. Edited by B. Eckmann. IV, 398 pages. 1969. DM 20,-

Vol. 81: J.-P. Eckmann et M. Guenin, Méthodes Algébriques en Mécanique Statistique. VI, 131 pages. 1969. DM 16,-

Vol. 82: J. Wloka, Grundräume und verallgemeinerte Funktionen. VIII, 131 Seiten. 1969. DM 16,-

Vol. 83: O. Zariski, An Introduction to the Theory of Algebraic Surfaces. IV, 100 pages. 1969. DM 16,-

Vol. 84: H. Lüneburg, Transitive Erweiterungen endlicher Permutationsgruppen. IV, 119 Seiten. 1969. DM 16,-

Vol. 85: P. Cartier et D. Foata, Problèmes combinatoires de commutation et réarrangements. IV, 88 pages. 1969. DM 16,-

Vol. 86: Category Theory, Homology Theory and their Applications I. Edited by P. Hilton. VI, 216 pages. 1969. DM 16,-

Vol. 87: M. Tierney, Categorical Constructions in Stable Homotopy Theory. IV, 65 pages. 1969. DM 16,-

Vol. 88: Séminaire de Probabilités III. IV, 229 pages. 1969. DM 18,-

Vol. 89: Probability and Information Theory. Edited by M. Behara, K. Krickeberg and J. Wolfowitz. IV, 256 pages. 1969. DM 18,-

Vol. 90: N. P. Bhatia and O. Hajek, Local Semi-Dynamical Systems. II, 157 pages. 1969. DM 16,-

Vol. 91: N. N. Janenko, Die Zwischenschrittmethode zur Lösung mehrdimensionaler Probleme der mathematischen Physik. VIII, 194 Seiten. 1969. DM 16,80

Vol. 92: Category Theory, Homology Theory and their Applications II. Edited by P. Hilton. V, 308 pages. 1969. DM 20,-

Vol. 93: K. R. Parthasarathy, Multipliers on Locally Compact Groups. III, 54 pages. 1969. DM 16,-

Vol. 94: M. Machover and J. Hirschfeld, Lectures on Non-Standard Analysis. VI, 79 pages. 1969. DM 16,-

Vol. 95: A. S. Troelstra, Principles of Intuitionism. II, 111 pages. 1969. DM 16,-

Vol. 96: H.-B. Brinkmann und D. Puppe, Abelsche und exakte Kategorien, Korrespondenzen. V, 141 Seiten. 1969. DM 16,-

Vol. 97: S. O. Chase and M. E. Sweedler, Hopf Algebras and Galois theory. II, 133 pages. 1969. DM 16,-

Vol. 98: M. Heins, Hardy Classes on Riemann Surfaces. III, 106 pages. 1969. DM 16,-

Vol. 99: Category Theory, Homology Theory and their Applications III. Edited by P. Hilton. IV, 489 pages. 1969. DM 24,-

Vol. 100: M. Artin and B. Mazur, Etale Homotopy. II, 196 Seiten. 1969. DM 16,-

Vol. 101: G. P. Szegö et G. Treccani, Semigruppi di Trasformazioni Multivoche. VI, 177 pages. 1969. DM 16,-

Vol. 102: F. Stummel, Rand- und Eigenwertaufgaben in Sobolewschen Räumen. VIII, 386 Seiten. 1969. DM 20,-

Vol. 103: Lectures in Modern Analysis and Applications I. Edited by C. T. Taam. VII, 162 pages. 1969. DM 16,-

Vol. 104: G. H. Pimbley, Jr., Eigenfunction Branches of Nonlinear Operators and their Bifurcations. II, 128 pages. 1969. DM 16,-

Vol. 105: R. Larsen, The Multiplier Problem. VII, 284 pages. 1969. DM 18,-

Vol. 106: Reports of the Midwest Category Seminar III. Edited by S. Mac Lane. III, 247 pages. 1969. DM 16,-

Vol. 107: A. Peyerimhoff, Lectures on Summability. III, 111 pages. 1969. DM 16,-

Vol. 108: Algebraic K-Theory and its Geometric Applications. Edited by R. M. F. Moss and C. B. Thomas. IV, 86 pages. 1969. DM 16,-

Vol. 109: Conference on the Numerical Solution of Differential Equations. Edited by J. Ll. Morris. VI, 275 pages. 1969. DM 18,-

Vol. 110: The Many Facets of Graph Theory. Edited by G. Chartrand and S. F. Kapoor. VIII, 290 pages. 1969. DM 18,-

Vol. 111: K. H. Mayer, Relationen zwischen charakteristischen Zahlen. III, 99 Seiten. 1969. DM 16,-

Vol. 112: Colloquium on Methods of Optimization. Edited by N. N. Moiseev. IV, 293 pages. 1970. DM 18.

Vol. 113: R. Wille, Kongruenzklassengeometrien. III, 99 Seiten. 1970. DM 16,-

Vol. 114: H. Jacquet and R. P. Langlands, Automorphic Forms on GL (2) VII, 548 pages. 1970. DM 24,-

Vol. 115: K. H. Roggenkamp and V. Huber-Dyson, Lattices over Orders I. XIX, 290 pages. 1970. DM 18,

Vol. 116: Séminaire Pierre Lelong (Analyse) Année 1969. IV, 195 pages. 1970. DM 16,-

Vol. 117: Y. Meyer, Nombres de Pisot, Nombres de Salem et Analyse Harmonique. 63 pages. 1970. DM 16,-

Vol. 118: Proceedings of the 15th Scandinavian Congress, Oslo 1968. Edited by K. E. Aubert and W. Ljunggren. IV, 162 pages. 1970. DM 16,-

Vol. 119: M. Raynaud, Faisceaux amples sur les schémas en groupes et les espaces homogènes. III, 219 pages. 1970. DM 16,-

Vol. 120: D. Siefkes, Büchi's Monadic Second Order Successor Arithmetic. XII, 130 Seiten. 1970. DM 16,-

Vol. 121: H. S. Bear, Lectures on Gleason Parts. III, 47 pages. 1970. DM 16,-

Vol. 122: H. Zieschang, E. Vogt und H.-D. Coldewey, Flächen und ebene diskontinuierliche Gruppen. VIII, 203 Seiten. 1970. DM 16,-

Vol. 123: A. V. Jategaonkar, Left Principal Ideal Rings. VI, 145 pages. 1970. DM 16,-

Vol. 124: Séminare de Probabilités IV. Edited by P. A. Meyer. IV, 282 pages. 1970. DM 20,-

Vol. 125: Symposium on Automatic Demonstration. V, 310 pages. 1970. DM 20,-

Vol. 126: P. Schapira, Théorie des Hyperfonctions. XI, 157 pages. 1970. DM 16,-

Vol. 127: I. Stewart, Lie Algebras. IV, 97 pages. 1970. DM 16,-

Vol. 128: M. Takesaki, Tomita's Theory of Modular Hilbert Algebras and its Applications. II, 123 pages. 1970. DM 16,-

Vol. 129: K. H. Hofmann, The Duality of Compact Semigroups and C*- Bigebras. XII, 142 pages. 1970. DM 16,-

Vol. 130: F. Lorenz, Quadratische Formen über Körpern. II, 77 Seiten. 1970. DM 16,-

Vol. 131: A Borel et al., Seminar on Algebraic Groups and Related Finite Groups. VII, 321 pages. 1970. DM 22,-

Vol. 132: Symposium on Optimization. III, 348 pages. 1970. DM 22,-

Vol. 133: F. Topsøe, Topology and Measure. XIV, 79 pages. 1970. DM 16,-

Vol. 134: L. Smith, Lectures on the Eilenberg-Moore Spectral Sequence. VII, 142 pages. 1970. DM 16,-

Vol. 135: W. Stoll, Value Distribution of Holomorphic Maps into Compact Complex Manifolds. II, 267 pages. 1970. DM 18,-

Vol. 136: M. Karoubi et al., Séminaire Heidelberg-Saarbrücken-Strasbourg sur la K-Théorie. IV, 264 pages. 1970. DM 18,-

Vol. 137: Reports of the Midwest Category Seminar IV. Edited by S. MacLane. III, 139 pages. 1970. DM 16,-

Vol. 138: D. Foata et M. Schützenberger, Théorie Géométrique des Polynômes Eulériens. V, 94 pages. 1970. DM 16,-

Vol. 139: A. Badrikian, Séminaire sur les Fonctions Aléatoires Linéaires et les Mesures Cylindriques. VII, 221 pages. 1970. DM 18,-

Vol. 140: Lectures in Modern Analysis and Applications II. Edited by C. T. Taam. VI, 119 pages. 1970. DM 16,-

Vol. 141: G. Jameson, Ordered Linear Spaces. XV, 194 pages. 1970. DM 16,-

Vol. 142: K. W. Roggenkamp, Lattices over Orders II. V, 388 pages. 1970. DM 22,-

Vol. 143: K. W. Gruenberg, Cohomological Topics in Group Theory. XIV, 275 pages. 1970. DM 20,-

Vol. 144: Seminar on Differential Equations and Dynamical Systems, II. Edited by J. A. Yorke. VIII, 268 pages. 1970. DM 20,-

Vol. 145: E. J. Dubuc, Kan Extensions in Enriched Category Theory. XVI, 173 pages. 1970. DM 16,-

Vol. 146: A. B. Altman and S. Kleiman, Introduction to Grothendieck Duality Theory. II, 192 pages. 1970. DM 18,

Vol. 147: D. E. Dobbs, Cech Cohomological Dimensions for Commutative Rings. VI, 176 pages. 1970. DM 16,-

Vol. 148: R. Azencott, Espaces de Poisson des Groupes Localement Compacts. IX, 141 pages. 1970. DM 16,-

Vol. 149: R. G. Swan and E. G. Evans, K-Theory of Finite Groups and Orders. IV, 237 pages. 1970. DM 20,-

Vol. 150: Heyer, Dualität lokalkompakter Gruppen. XIII, 372 Seiten. 1970. DM 20,-

Vol. 151: M. Demazure et A. Grothendieck, Schémas en Groupes I. (SGA 3). XV, 562 pages. 1970. DM 24,

Vol. 152: M. Demazure et A. Grothendieck, Schémas en Groupes II. (SGA 3). IX, 654 pages. 1970. DM 24,-

Vol. 153: M. Demazure et A. Grothendieck, Schémas en Groupes III. (SGA 3). VIII, 529 pages. 1970. DM 24,-

Vol. 154: A. Lascoux et M. Berger, Variétés Kähleriennes Compactes. VII, 83 pages. 1970. DM 16,-

Vol. 155: Several Complex Variables I, Maryland 1970. Edited by J. Horváth. IV, 214 pages. 1970. DM 18,-

Vol. 156: R. Hartshorne, Ample Subvarieties of Algebraic Varieties. XIV, 256 pages. 1970. DM 20,-

Vol. 157: T. tom Dieck, K. H. Kamps und D. Puppe, Homotopietheorie. VI, 265 Seiten. 1970. DM 20,-

Vol. 158: T. G. Ostrom, Finite Translation Planes. IV, 112 pages. 1970. DM 16,-

Vol. 159: R. Ansorge und R. Hass. Konvergenz von Differenzenverfahren für lineare und nichtlineare Anfangswertaufgaben. VIII, 145 Seiten. 1970. DM 16,-

Vol. 160: L. Sucheston, Constributions to Ergodic Theory and Probability. VII, 277 pages. 1970. DM 20,-

Vol. 161: J. Stasheff, H-Spaces from a Homotopy Point of View. VI, 95 pages. 1970. DM 16,-

Vol. 162: Harish-Chandra and van Dijk, Harmonic Analysis on Reductive p-adic Groups. IV, 125 pages. 1970. DM 16,-

Vol. 163: P. Deligne, Equations Différentielles à Points Singuliers Reguliers. III, 133 pages. 1970. DM 16,-

Vol. 164: J. P. Ferrier, Seminaire sur les Algebres Complétes. II, 69 pages. 1970. DM 16,-

Vol. 165: J. M. Cohen, Stable Homotopy. V, 194 pages. 1970. DM 16,-

Vol. 166: A. J. Silberger, PGL_2 over the p-adics: its Representations, Spherical Functions, and Fourier Analysis. VII, 202 pages. 1970. DM 18,-

Vol. 167: Lavrentiev, Romanov and Vasiliev, Multidimensional Inverse Problems for Differential Equations. V, 59 pages. 1970. DM 16,-

Vol. 168: F. P. Peterson, The Steenrod Algebra and its Applications: A conference to Celebrate N. E. Steenrod's Sixtieth Birthday. VII, 317 pages. 1970. DM 22,-

Vol. 169: M. Raynaud, Anneaux Locaux Henséliens. V, 129 pages. 1970. DM 16,-

Vol. 170: Lectures in Modern Analysis and Applications III. Edited by C. T. Taam. VI, 213 pages. 1970. DM 18,-

Vol. 171: Set-Valued Mappings, Selections and Topological Properties of 2^X. Edited by W. M. Fleischman. X, 110 pages. 1970. DM 16,-

Vol. 172: Y.-T. Siu and G. Trautmann, Gap-Sheaves and Extension of Coherent Analytic Subsheaves. V, 172 pages. 1971. DM 16,-

Vol. 173: J. N. Mordeson and B. Vinograde, Structure of Arbitrary Purely Inseparable Extension Fields. IV, 138 pages. 1970. DM 16,-

Vol. 174: B. Iversen, Linear Determinants with Applications to the Picard Scheme of a Family of Algebraic Curves. VI, 69 pages. 1970. DM 16,-

Vol. 175: M. Brelot, On Topologies and Boundaries in Potential Theory. VI, 176 pages. 1971. DM 18,-

Vol. 176: H. Popp, Fundamentalgruppen algebraischer Mannigfaltigkeiten. IV, 154 Seiten. 1970. DM 16,

Vol. 177: J. Lambek, Torsion Theories, Additive Semantics and Rings of Quotients. VI, 94 pages. 1971. DM 16,-

Vol. 178: Th. Bröcker und T. tom Dieck, Kobordismentheorie. XVI, 191 Seiten. 1970. DM 18,-

Vol. 179: Seminaire Bourbaki - vol. 1968/69. Exposés 347-363. IV, 295 pages. 1971. DM 22,-

Vol. 180: Séminaire Bourbaki - vol. 1969/70. Exposés 364-381. IV, 310 pages. 1971. DM 22,-

Vol. 181: F. DeMeyer and E. Ingraham, Separable Algebras over Commutative Rings. V, 157 pages. 1971. DM 16.